Introduction to Web Mapping

T0188251

Introduction to Web Mapping

Michael Dorman

CRC Press
Taylor & Francis Group
Boca Raton London New York

CRC Press is an imprint of the
Taylor & Francis Group, an **informa** business

A CHAPMAN & HALL BOOK

CRC Press
Taylor & Francis Group
6000 Broken Sound Parkway NW, Suite 300
Boca Raton, FL 33487-2742

© 2020 by Taylor & Francis Group, LLC
CRC Press is an imprint of Taylor & Francis Group, an Informa business

No claim to original U.S. Government works

Printed on acid-free paper

International Standard Book Number-13: 978-0-367-37139-5 (Hardback)
978-0-367-86118-6 (Paperback)

Visit the Taylor & Francis Web site at
http://www.taylorandfrancis.com

and the CRC Press Web site at
http://www.crcpress.com

To my daughter Ariel,

who inspires me to explore new things

Contents

Preface

0.1 What is web mapping?

A **web map**[1] is an interactive display of geographic information, in the form of a web page, that you can use to tell stories and answer questions. In the past, most digital geographic information was confined to specialized software on desktop PCs and could not be easily shared. With the advent of web mapping, geographical information can be shared, visualized, and edited in the browser. The most important advantage to this is accessibility: a web map, just like any website, can be reached by anyone from any device that has an internet browser and an internet connection.

Web maps are **interactive**. The term interactive implies that the viewer can interact with the map. This can mean selecting different map data layers or features to view, zooming into a particular part of the map that you are interested in, inspecting feature properties, editing existing content, or submitting new content, and so on.

Web maps are also said to be powered by the web, rather than just digital maps on the web. This means that the map is usually not self-contained; in other words, it depends on the internet. At least some of the content displayed on a web maps is usually loaded from other locations on the web, such as a tile server (Section 6.5.6.2).

Web maps are useful for various purposes, such as data visualization in journalism (and elsewhere), displaying real-time spatial data, powering spatial queries in online catalogs and search tools, providing computational tools, reporting, and collaborative mapping. Some examples of web maps used for different purposes are given in Table 0.1. All of these are built with **JavaScript**.

TABLE 0.1: Examples of web maps for different purposes

Purpose	Example	Notes
Visualization and journalism	Global Migration[2]	
	Ship Traffic[3]	
	Israel Municipalities[4]	
	NYC Planning[5]	
Real-time information	Earth Weather[6]	Figure 1
	Real-Time Transport Location[7]	

---

[1]https://en.wikipedia.org/wiki/Web_mapping
[2]http://metrocosm.com/global-migration-map.html
[3]https://www.shipmap.org/
[4]http://mindthemap.info/mtm/
[5]https://zola.planning.nyc.gov/
[6]https://earth.nullschool.net/
[7]https://traze.app/

Purpose	Example	Notes
	Real-Time Flight Locations[8]	
	Stuff in Space[9]	
	Global Forest Watch[10]	
Catalog and search	Earth Data Search[11]	
Computational tools	Google Maps[12]	
	SunCalc[13]	
	geojson.io[14]	see Section 7.4.1
	mapshaper[15]	see Section 7.4.2
	Route Planner[16]	
Reporting and collaboration	OpenStreetMap[17]	
	Falling Fruit[18]	

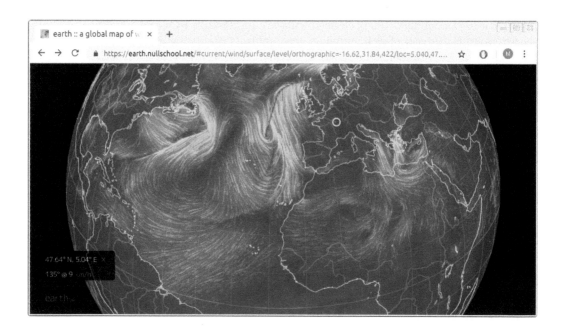

FIGURE 1: Real-time earth weather visualization on `https://earth.nullschool.net`

---

[8] `https://www.flightradar24.com/`
[9] `http://stuffin.space/`
[10] `https://www.globalforestwatch.org/`
[11] `https://search.earthdata.nasa.gov/`
[12] `https://www.google.com/maps`
[13] `http://suncalc.net/`
[14] `http://geojson.io/`
[15] `http://mapshaper.org/`
[16] `https://www.outdooractive.com/en/routeplanner/`
[17] `http://www.openstreetmap.org`
[18] `https://fallingfruit.org/`

0.2 What is JavaScript?

JavaScript is a programming language, primarily used to control interactive behavior in web pages. Alongside **HTML** and **CSS**, JavaScript is one of the three core technologies of the web. JavaScript is the only programming language that can run in the **web browser**, and it is used by ~95% of websites. The importance of the internet thus makes JavaScript one of the most popular programming[19] languages (Figure 2).

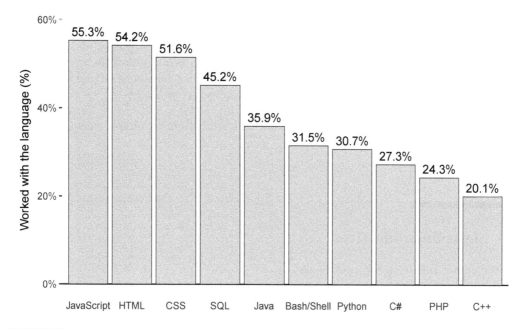

FIGURE 2: Programming language popularity, based on the StackOverflow survey for 2018

JavaScript and web development are huge topics. In this book, we are going to learn JavaScript from the specific point of view of web mapping.

0.3 Why use JavaScript for web mapping?

The availability of code-free, graphical interfaces for making sophisticated web maps remains an ongoing goal for the web-mapping community. Several platforms now provide straightforward interactive interfaces for building and publishing web maps and web-mapping applications, such as:

- CARTO Builder[20]

---

[19]Based on the StackOverflow survey for 2018 (`https://stackoverflow.blog/2018/05/30/public-data-release-of-stack-overflows-2018-developer-survey/`).

[20]`https://carto.com/builder/`

- Mapbox's Studio[21]
- ESRI's Configurable Apps[22]
- qgis2web Plugin for QGIS[23]

A more customizable, programmatic approach for building web maps is available through scripting languages such as **R** and **Python**. Both languages have libraries that give the ability to build web maps, using few lines of code and incorporating data from the user environment:

- `leaflet`[24] and `mapview`[25] (**R** packages)
- `folium`[26] (**Python** library)

All of the above eventually build HTML, CSS, and JavaScript code, with varying degrees of flexibility and customization.

Nevertheless, proficiency in the fundamental web technologies HTML, CSS, and JavaScript eventually allows web cartographers to control all low-level properties of the web map they are building. That way, the user experience of web maps can be enhanced and customized beyond what is provided by either of the above "indirect," or high-level, approaches. For example, the **Leaflet** web mapping JavaScript library has a wide range of plugins and extensions[27], mostly unavailable in external tools and libraries unless using it directly, through JavaScript coding.

0.4   Learning objectives

By the end of this book, you will be able to:

- Build and publish basic websites and web maps
- Use JavaScript to add interactive behavior in your maps
- Connect you web map to a database to display large amounts of data
- Include client-side geoprocessing functionality in your web map
- Gain an understanding on how the web works, and a starting point for learning more

0.5   Software

The field of web mapping, much like web development as a whole, is rapidly changing. The book thus intends to emphasize established technologies, libraries, and principles which are unlikely to go away soon. These include HTML, CSS, JavaScript, jQuery, Leaflet, GeoJSON, Ajax, and PostGIS (Figure 3).

---

[21]https://www.mapbox.com/mapbox-studio/
[22]http://www.esri.com/software/configurable-apps
[23]https://github.com/tomchadwin/qgis2web
[24]https://rstudio.github.io/leaflet/
[25]https://r-spatial.github.io/mapview/
[26]http://python-visualization.github.io/folium/
[27]https://leafletjs.com/plugins.html

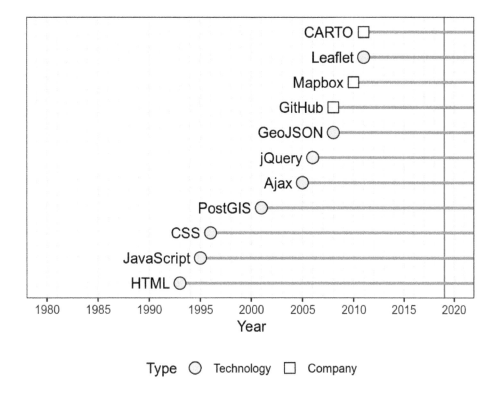

FIGURE 3: Initial release time of technologies and platforms used in the book (release year data from Wikipedia)

Throughout the book, we are going to use the following software components, which are freely available on the web:

- A **plain text editor**, such as:
    - Notepad++[28]
    - Sublime Text[29]
    - VS Code[30]
    - Atom[31]
    - Brackets[32]
- A **modern web browser**[33]:
    - Chrome[34] is recommended, because its developer tools are demonstrated in the examples and figures throughout the book
    - Firefox[35] is also a good option

---

[28]https://notepad-plus-plus.org/
[29]https://www.sublimetext.com/
[30]https://code.visualstudio.com/
[31]https://atom.io/
[32]http://brackets.io/
[33]There is no clear definition of the term *modern web browser*, but older versions of Internet Explorer are, arguably, not considered as modern browsers and thus not recommended to illustrate web development (https://teamtreehouse.com/community/what-is-a-modern-browser).
[34]https://www.google.com/chrome/
[35]https://www.mozilla.org/en-US/firefox/

- **Web developer tools** (built into Chrome and Firefox)
- A **local web server** (built into Python, other options are available)

We will introduce and demonstrate five open-source JavaScript **libraries**, as listed on Table 0.2. The newest versions of these libraries, at the time of writing (May 2019), are being used throughout the book. The online version of the book (Section 0.7) includes copies of those specific library versions, which you can download to make sure the examples from the book are reproducible, even if newer versions of the libraries will introduce conflicting changes in the future.

TABLE 0.2: JavaScript library versions used in the book

Library	Version	Released	URL
jQuery	3.4.1	2019-05-01	`https://jquery.com/`
Leaflet	1.5.1	2019-05-08	`https://leafletjs.com/`
Turf.js	5.1.6	2017-12-10	`http://turfjs.org/`
Leaflet.heat	0.2.0	2015-10-26	`https://github.com/Leaflet/Leaflet.heat`
Leaflet.draw	1.0.4	2018-10-24	`https://github.com/Leaflet/Leaflet.draw`

While emphasizing open-source solutions, we will also demonstrate the commercial **services** GitHub[36] and CARTO[37], chosen for their simplicity. Free and open-source alternatives giving the same functionality will be mentioned for those willing to explore more complicated options. Several datasets from the book are also demonstrated using the QGIS[38] software (Figures 8.3, 9.3), but it is not essential to be familiar with QGIS, or any other geographic information system (GIS) software, to follow the material.

0.6   Background knowledge

The book can be used as primary text for a web-mapping course in geography departments, or by anyone who is interested in the topic. Each of the thirteen chapters is designed to be covered in a three-hour lecture, or through self-study. Short questions and exercises are given throughout the chapters to demonstrate the material from different angles and facilitate understanding. There are also concluding exercises with complete solutions (Appendix C) at the end of Chapters 3–4 and 6–12.

Familiarity with basic concepts of geographic data and GIS (coordinate systems, projections, spatial layer file formats, etc.) is necessary for deeper understanding of some of the topics in the book. Readers who are not familiar with GIS can skip the theoretic considerations and still follow the material from the technical point of view.

The book assumes no background knowledge in programming or web technologies (HTML, CSS, JavaScript), going through all necessary material from the beginning. Previous experience with programming (e.g., using Python or R) and using databases (SQL) is beneficial but not required.

---

[36]`https://github.com/`
[37]`https://carto.com/`
[38]`https://qgis.org/`

0.7 Online version

The online version of this book is freely available, with agreement from the publisher, at the following addresses:

- `http://geobgu.xyz/web-mapping/`
- `https://web-mapping.surge.sh/`

In addition to the content of the printed version, the online version includes live versions and downloadable code for all ninety-plus complete examples (Appendix B) and exercise solutions (Appendix C), as well as an additional appendix with instructions for setting up an SQL API (see Section 9.2.2).

0.8 Acknowledgments

I thank the creators and contributors of the open-source tools used in this book, namely jQuery, Leaflet, Turf.js, Leaflet.heat, and Leaflet.draw (Table 0.2). Vladimir Agafonkin, the creator of the Leaflet web-mapping library, deserves special mention here. Leaflet took open-source web mapping to a new level: for me to write this book, and for readers to enter the field of web mapping, would have been much more difficult without Leaflet. I am also grateful to Yihui Xie, whose `bookdown` (Xie, 2018, 2016) R package (R Core Team, 2018) greatly facilitated the technical process of writing the book.

Figure 6.1 was prepared using R packages `sf` (Pebesma, 2018) and `rworldmap` (South, 2011). Figure 7.1 and the images inside Tables 7.2–7.6 were also prepared using package `sf`. Figure 8.4 was prepared using R package `RColorBrewer` (Neuwirth, 2014). Figures 8.3 and 9.3 were prepared using QGIS (QGIS Development Team, 2018). The sample database shown in Figure 9.1 was prepared using data from R package `nycflights13` (Wickham, 2018). Figures 6.2 and 6.3 display images of OpenStreetMap tiles (© OpenStreetMap contributors).

I would like to thank Rob Calver and Lara Spieker from CRC Press for working with me and supporting this project, starting from a talk with Rob at a conference in Brussels in summer 2017, going through project initiation, the professional review process, and up to finalizing the manuscript and bringing it to press. I also thank Annie Sophia, Vaishali Singh and Jyotsna Jangra for the professional proofreading and production of the book. Shashi Kumar and John Kimmel from CRC are acknowledged for their assistance in technical issues.

This book is based on the materials of the course *Introduction to JavaScript for Web Mapping*, given by the Department of Geography, Ben-Gurion University of the Negev, in Spring 2018 and 2019. I am grateful to Prof. Tal Svoray and Prof. Itai Kloog from the department for their guidance and encouragement during the development of course materials, and to the students for valuable feedback. I thank Shai Sussman and three anonymous reviewers for comments that greatly helped improve the manuscript.

Several figures in the book include icons from external sources: "computer," "web page," and "car" icons by Freepik[39], "folder" and "database" icons by Smashicons[40], "gears" icon by Good Ware[41], from `https://www.flaticon.com/`, licensed by CC 3.0 BY[42].

---

[39]`http://www.freepik.com`
[40]`https://www.flaticon.com/authors/smashicons`
[41]`https://www.flaticon.com/authors/good-ware`
[42]`http://creativecommons.org/licenses/by/3.0/`

Author

Michael Dorman is a programmer (since 2016) and lecturer (since 2013) at the Department of Geography and Environmental Development, Ben-Gurion University of the Negev. He is working with researchers and students in the department in developing computational workflows such as data processing, spatial analysis, geostatistics, development of web applications and web maps, etc., mostly through programming in R, JavaScript, and Python. In 2018, he developed and taught a course named Introduction to JavaScript for Web Mapping, introducing web technologies and web mapping to undergraduate geography students specializing in Geographic Information Systems (GIS). The course materials served as a foundation for this book. Michael holds a Ph.D. in Geography and a M.Sc. in Life Sciences from the Ben-Gurion University of the Negev, and a B.Sc. in Plant Sciences in Agriculture from The Hebrew University of Jerusalem. He previously published the book *Learning R for Geospatial Analysis* (Packt Publishing, 2014) and authored or coauthored twenty-seven papers in the scientific literature.

Part I

Introduction to Web Technologies

1

HTML

1.1 Introduction

In this chapter, we introduce the most basic and fundamental component of web technologies: **HTML**. As we will see, HTML is a data format used to encode the contents and structure of web pages. HTML is usually stored in plain text files with the `.html` file extension. Every time one accesses a web page, using a web browser, the respective HTML file for that web page is transferred and decoded into the visual image that we see on screen. Simple web pages, such as the ones we build in this chapter, are composed of nothing but HTML code contained in a single `.html` file.

Starting from this chapter and onward, we are going to present **computer code** examples. Some examples are short, separate pieces of code used to illustrate an idea or concept. Other examples include the complete source code of a web page, which you can open and display in the browser, as well as modify and experiment with. The way that each of the complete code examples will appear when opened with the browser is shown in a separate figure, such as in Figure 1.1. As mentioned in Section 0.7, the online version of this book contains live versions of all ninety-plus complete examples (Appendices B–C), as well as a downloadable folder with all code files to experiment with the examples on your own computer.

Learning programming requires a lot of practice, so it is highly recommended to open the examples on your computer as you go along through the book. Better yet, you can modify the code of each example and observe the way that the displayed result changes, to make sure you understand what is the purpose of each code component. For instance, the first example (Figure 1.1) displays a simple web page with one heading and one paragraph—you can try to modify its source code (see Section 1.4 to learn how) to change the contents of the heading and/or paragraph, to add a second paragraph below the first one, and so on.

Chapter 2 in *Introduction to Data Technologies* (Murrell, 2009) gives a gentle and gradual introduction to HTML as well as the practice of writing computer code[1]. It is a highly recommended complementary reading to the present chapter, especially for readers who are new to computer programming.

---

[1]The book has a freely available PDF version here: `https://www.stat.auckland.ac.nz/~paul/ItDT/`.

1.2   How do people access the web?

1.2.1   Web browsers

People access the web using software known as a **web browser**[2]. Popular examples of web browsers are listed in Table 1.1.

TABLE 1.1: Popular web browsers

Browser	URL
Chrome	`https://www.google.com/chrome/`
Firefox	`https://www.mozilla.org/en-US/firefox/`
Edge	`https://www.microsoft.com/en-us/windows/microsoft-edge`
Internet Explorer	`https://www.microsoft.com/en-us/download/internet-explorer.aspx`
Safari	`https://www.apple.com/lae/safari/`

In order to view a web page, users might:

- Type a URL into the address bar of the browser
- Follow a link from another site
- Use a bookmark

1.2.2   Web servers

When you ask your browser for a web page, typing a **URL**[3] such as `https://www.google.com` in the address bar, the request is sent across the internet to a special computer known as a **web server** which hosts the website. Web servers are special computers that are constantly connected to the internet, and are optimized to send web pages out to people who request them. Your computer, the **client**, receives the file and renders the web page you ultimately see on screen. We will discuss web servers and server-client communication in Chapter 5.

When you are looking at a website, it is most likely that your browser will be receiving **HTML** and **CSS** documents from the web server that hosts the site. The web browser interprets the HTML and CSS code to create the page that you see. We will learn about HTML in Chapter 1 (this chapter) and about CSS in Chapter 2.

Most web pages also send **JavaScript** code to your browser to make the page interactive. The browser runs the JavaScript code, on page load and/or later on while the user interacts with the web page. The JavaScript code can modify the content of the page. We will introduce JavaScript in Chapters 3–4.

---

[2]`https://en.wikipedia.org/wiki/Web_browser`
[3]`https://en.wikipedia.org/wiki/URL`

1.3 Web pages

At the most basic level, a **web page** is a plain text document containing HTML code. This book comes with several examples of complete web pages. The examples are listed in Appendices B–C. They can be viewed and/or downloaded from the online version of this book (Section 0.7).

The first example, `example-01-01.html`, is a *minimal* HTML document. When opening this file in the browser, a minimal web page is displayed (Figure 1.1).

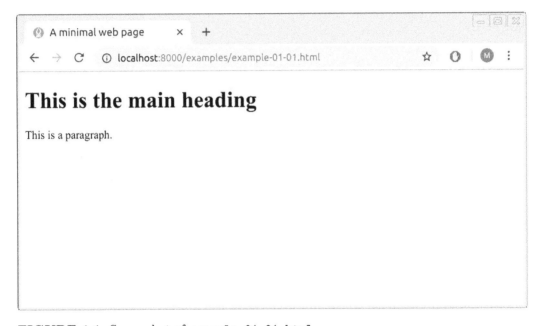

FIGURE 1.1: Screenshot of `example-01-01.html`

Here is the source code you should see when opening the file `example-01-01.html` in a plain text editor (Figure 1.2), or in the source code tab in the browser (Figure 1.3):

```
<!DOCTYPE html>
<html>
    <head>
        <title>A minimal web page</title>
    </head>
    <body>
        <h1>This is the main heading</h1>
        <p>This is a paragraph.</p>
    </body>
</html>
```

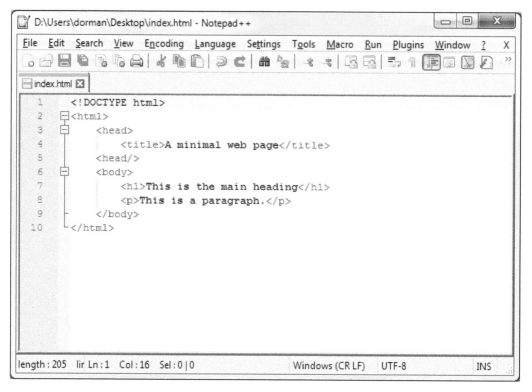

FIGURE 1.2: HTML document source code viewed in a text editor (Notepad++)

- Download the book materials from the online version (Section 0.7).
- Open the first example, a file named `example-01-01.html`, in a web browser such as Chrome, by double-clicking on the file, or by clicking with the right mouse button and selecting **Open with...** then choosing a web browser program[4]. The file `example-01-01.html` is a minimal HTML document, as shown on the left pane of Figure 1.3.
- Press **Ctrl+U** (in Chrome) to open a new tab with the source code that was used to create the page, as shown on the right pane of Figure 1.3.

The source code comprises the contents of an HTML document. The source code is sent to the browser, then processed to produce the display shown in Figure 1.1.

The `example-01-01.html` file contains a minimal web page, built using a single (short) HTML document. As we shall see throughout the book, more complicated web pages also include CSS and/or JavaScript code, possibly in separate files linked to the main document. Each of these three languages has a different role when building the web page you see in the browser:

---

[4]Opening an HTML document by double clicking on it is quick and simple, yet it is not suitable for displaying more complex web page components, which we are going to use starting from Chapter 7. In Chapter 5 we are going to learn the "right" way to view a web page we are developing—using a web server.

FIGURE 1.3: HTML document (left) and its source code (right)

- **HTML** (`.html`)—Determines page **contents**
- **CSS** (`.css`)—Determines presentation **style**
- **JavaScript** (`.js`)—Determines interactive **behavior**

1.4   Text editors

HTML, CSS, and JavaScript code, like any other computer code, is **plain text**[5] stored in text files. To edit them, you need to use a plain text editor. The simplest option is **Notepad++**[6]. There are also more advanced editors such as **Sublime Text**[7]. The more advanced editors contain additional features for easier text editing, such as shortcuts, highlighted syntax, marked matching brackets, etc. You can use any plain text editor you prefer[8].

- Open a plain text editor, such as **Notepad++**.
- Copy the HTML code section given above (Section 1.3) and paste it into a blank text document (Figure 1.2).
- Click **Save As...** and save the document to a file named `index.html`. If you are using Notepad++, make sure you choose **Save as type...** and select the *Hyper Text Markup Language (HTML)* file format.
- Go to the location where you saved the file on your computer and double click on the file.
- The browser should now open and display the minimal web page!
- Go back to the text editor, locate the text `This is a paragraph.`, replace it with any other text of your choice and save the document.
- Refresh the browser—you should see the new text displayed on the web page!

---

[5]https://en.wikipedia.org/wiki/Plain_text
[6]https://notepad-plus-plus.org/
[7]https://www.sublimetext.com/
[8]See Section 0.5 for a list of recommended plain text editors.

1.5   What is HTML?

1.5.1   Overview

Hypertext Markup Language (HTML)[9] is the language that describes the contents and structure of web pages. Most web pages today are composed of more than just HTML, but simple web pages—such as `example-01-01.html` and the other examples we create in this chapter—can be made with HTML alone.

HTML code consists of HTML **elements**[10]. An HTML element contains text and/or other elements. This makes HTML code *hierarchical*. An HTML element consists of a start **tag**, followed by the element content, followed by an end tag. A start tag is of the form `<elementName>` and an end tag is of the form `</elementName>`. The start and end tags contain the element name (`elementName`).

The following example shows a `<title>` element; the start tag is `<title>`, the end tag is `</title>`, and the contents is the text `Web Mapping`:

```
<title>Web Mapping</title>
```

Table 1.2 summarizes the basic components of an HTML element.

TABLE 1.2: HTML element structure

Component	Example
HTML element	`<title>Web Mapping</title>`
Start tag	`<title>Web Mapping</title>`
Element name	`<title>Web Mapping</title>`
Element contents	`<title>Web Mapping</title>`
End tag	`<title>Web Mapping</title>`

Some HTML elements are *empty*, which means that they consist of only a start tag, with no contents and no end tag. The following code shows an `<hr>` element, which is an example of an empty element:

```
<hr>
```

An element may have one or more **attributes**[11]. Attributes appear inside the start tag and are of the form `attributeName="attributeValue"`. The following code section shows an example of an `<img>` element, with an attribute called `src`. The value of the attribute in this example is `"images/leaflet.png"`. Note that `<img>`, like `<hr>`, is an empty HTML element, which is why it does not have an end tag.

---

[9]https://en.wikipedia.org/wiki/HTML
[10]https://en.wikipedia.org/wiki/HTML_element
[11]https://en.wikipedia.org/wiki/HTML_attribute

```
<img src="images/leaflet.png">
```

Table 1.3 summarizes the components of an HTML element with an attribute.

TABLE 1.3: HTML element attribute structure

Component	Example
HTML element	`<img src="images/leaflet.png">`
Element name	`<img src="images/leaflet.png">`
Attribute	`<img src="images/leaflet.png">`
Attribute name	`<img src="images/leaflet.png">`
Attribute value	`<img src="images/leaflet.png">`

There can be more than one attribute for an element, in which case they are separated by spaces. For example, the following `<img>` element has two attributes, `src` and `width`:

```
<img src="images/leaflet.png" width="300px">
```

It is important to note that there is a fixed set of valid HTML elements (see below), and each element has its own set of possible attributes. Moreover, some attributes are *required* while others are *optional*. For example, the `src` attribute is required for the `<img>` element, but irrelevant for the `<title>` element. As we will see shortly, there are also rules regarding the elements that another element can contain. (Don't worry about the meaning of the element and attribute names we mentioned just yet, we will cover this shortly in Section 1.6.)

As for the entire document structure, an HTML document must include a `DOCTYPE` declaration and a single `<html>` element. Within the `<html>` element, there must be a single `<head>` element and a single `<body>` element. Within the `<head>` element there must be a `<title>` element. This leads us to the minimal HTML code shown below:

```
<!DOCTYPE html>
<html>
    <head>
        <title></title>
    </head>
    <body>
    </body>
</html>
```

Technically, everything except for the `DOCTYPE` declaration is optional[12] since in most cases the browser can automatically fill the missing parts. For clarity, most websites nevertheless include the above minimal structure, and so will we.

As mentioned above, the primary role of HTML code is to specify the contents of a web page. The type of elements being used and their ordering determine the structure of information that is being displayed in the browser.

---

[12]https://google.github.io/styleguide/htmlcssguide.html#Optional_Tags

1.5.2 HTML comments

It is good practice to keep **comments** in our code, so that we can remember our intentions later on, and so that other people reading our code can understand what we did more easily. In HTML, comments are written as follows:

```
<!-- This is a comment -->
```

Anything between the start `<!--` and end `-->`, including HTML tags, is completely ignored by the computer. It is only there to pass messages to a human reader of the code.

1.5.3 Block vs. inline

While learning about the various HTML elements, it is important to keep in mind that HTML elements are divided into two general types of behaviors[13]:

- **Block-level** elements
- **Inline** elements

A block-level element, or simply a **block** element, is like a paragraph. Block elements always start on a new line in the browser window (Figure 1.4). Examples of block elements include:

- Headings (`<h1>`)
- Paragraphs (`<p>`)
- Bullet-point lists (`<ul>`)
- Numbered lists (`<li>`)

It is helpful to imagine block elements as horizontal boxes. Box width is determined by the width of the browser window, so that the box fills the entire available space. Box height is determined by the amount of content. For example, a paragraph fills the entire available page width, with variable height depending on the amount of text. (This is the default behavior; in Chapter 2 we will see that the height and width can be modified using CSS.)

An **inline** element is like a word within a paragraph. It is a small component that is arranged with other components inside a container. Inline elements appear on the same line as their neighboring elements (Figure 1.4). Examples of inline elements include:

- Links (`<a>`)
- Bold text (`<b>`)
- Italic text (`<i>`)
- Images (`<img>`)

1.6 Common HTML elements

This section briefly describes the important behavior, attributes, and rules for each of the common HTML elements. We will use most of these elements throughout the book, so it

---

[13]https://developer.mozilla.org/en-US/docs/Web/HTML/Block-level_elements#Block-level_vs._inline

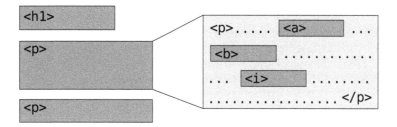

FIGURE 1.4: Block vs. inline HTML elements

is important to be familiar with them from the start. You don't need to remember how to use each element—you can always come back to this section later on. Keep in mind that the HTML elements we are going to cover in this chapter are just the most common ones. HTML defines a lot of other element types that we will not use in the book[14].

For convenience, the HTML elements we will cover will be divided into three types according to their *role* (Table 1.4) in determining page contents and structure. Other than elements setting the basic document **structure**, there are elements giving general **information** about the page (mainly inside the **<head>** element) and elements giving the actual **content** shown on screen (inside the **<body>** element). In the following Sections 1.6.1–1.6.12, we are going to cover the listed elements in the same order as given in Table 1.4.

TABLE 1.4: Common HTML elements

Role	Element	Description
Document structure	`<!DOCTYPE>`	Document type declaration
	`<html>`	Document
	`<head>`	General information
	`<body>`	Content
General information	`<title>`	Title
	`<meta>`	Metadata
	`<style>`	Embedded CSS code
	`<link>`	External CSS code
	`<script>`	JavaScript code
Content	`<h1>, <h2>, ..., <h6>`	Headings
	`<p>`	Paragraphs
	`<b>, <i>, <sup>, <sub>, <pre>`	Font types
	` , <hr>`	Spacing
	`<ul>, <ol>, <li>`	Lists
	`<a>`	Links
	`<img>`	Images
	`<table>, <th>, <tr>, <td>`	Tables
	`<div>, <span>`	Grouping
	`<input>, <select>, <option>`	User input

---

[14]A list of HTML elements can be found in various online resources, such as the *HTML elements reference* (https://developer.mozilla.org/en-US/docs/Web/HTML/Element) by Mozilla.

1.6.1   Structure

1.6.1.1   Overview

The `<!DOCTYPE>` declaration specifies the version of HTML that the document is written in. It is followed by the `<html>`, `<body>`, and `<head>` elements, which determine the top-level division of the HTML document into two components: general information about the page (inside the `<head>`) and the actual content (inside the `<body>`).

1.6.1.2   `<!DOCTYPE>`

The `<!DOCTYPE>` declaration must be the first thing in the HTML document, before the `<html>` tag. The `<!DOCTYPE>` declaration is in fact not an HTML tag. It is an instruction to the browser, telling it what version[15] of HTML the page is written in. In this book we will use the following declaration, which specifies we are using the **HTML5** version of HTML:

```
<!DOCTYPE html>
```

1.6.1.3   `<html>`

The opening `<html>` tag indicates that anything between it and a closing `</html>` tag is HTML code. The `<html>` element must contain exactly one `<head>` element followed by exactly one `<body>` element.

1.6.1.4   `<head>`

The `<head>` element contains information about the page, rather than information that is shown within the main part of the browser window.

1.6.1.5   `<body>`

Everything inside the `<body>` element is actually *displayed* inside the browser window. This is where page contents are specified.

Combining the `<!DOCTYPE>` declaration and the three structural HTML elements, we get the following "template" of a minimal HTML page, which we have already seen above:

```
<!DOCTYPE html>
<html>
    <head>
        <!-- General information goes here -->
    </head>
    <body>
        <!-- Page content goes here -->
    </body>
</html>
```

---

[15]https://en.wikipedia.org/wiki/HTML#HTML_versions_timeline

The other elements that we will learn about appear within the `<head>` or within the `<body>` element. The `<head>` element commonly contains the following elements:

* `<title>` for specifying page **title**
* `<meta>` elements for specifying page **metadata**
* `<style>` and `<link>` elements for loading **CSS** code
* `<script>` elements for loading **JavaScript** code

The `<body>` element contains mostly elements related to contents, such as paragraphs (`<p>`), lists (`<ol>`), images (`<img>`), and so on. In addition, the `<body>` can also contain `<script>` elements for JavaScript code, just like the `<head>`. Thus JavaScript code can be placed in the `<body>`, or in the `<head>` section of an HTML document, or in both. We are going to encounter both options for placing JavaScript code later on in the book (Section 4.5.1).

The following Sections 1.6.2–1.6.3 provide more details on the five elements that commonly occur in the `<head>` element. Then, Sections 1.6.4–1.6.12 describe elements that are found in the `<body>` element.

1.6.2 Title and metadata

1.6.2.1 `<title>`

The contents of the `<title>` element specify the page **title**. The title is either shown in the top of the browser window, above where you usually type in the URL of the page you want to visit, or on the tab for that page. The `<title>` element must be within the `<head>` element and must only contain text. For example, the `<title>` element in the HTML document of the online version of this book is:

```
<title>Introduction to Web Mapping</title>
```

* Open a web page of your choice in the Chrome browser.
* Press **Ctrl | U** to open the HTML source code (Figure 1.3).
* Try to locate the `<title>` element (or use **Ctrl+F** to search), and compare its contents with the title shown in the browser window.

1.6.2.2 `<meta>`

The `<meta>` element contains information about the web page, or its **metadata**. The `<meta>` element is typically used to specify page description, keywords, the name of the document author, last modified date, and other general information. The metadata may be used by the browser and by search engines to optimize display and indexing of the page, respectively.

The `<meta>` element is an empty element. It typically uses the **name** and **content** attributes to carry the metadata, or the **charset** attribute to specify **character encoding**[16] for the

---

[16]https://en.wikipedia.org/wiki/Character_encodings_in_HTML

document. In the following example we see a `<head>` element containing several `<meta>` elements, specifying various metadata items:

```
<head>
    <meta charset="UTF-8">
    <meta name="description" content="Free Web tutorials">
    <meta name="keywords" content="HTML,CSS,XML,JavaScript">
    <meta name="author" content="John Doe">
    <meta name="viewport" content="width=device-width, initial-scale=1.0">
</head>
```

For example, the `<meta>` element with `name="viewport"` specifies web page display instructions, which is useful to make sure the web page is correctly scaled across different devices. We are going to use this to disable unwanted scaling of Leaflet web maps on mobile devices (Section 6.5.7).

1.6.3   Styling and scripts

1.6.3.1   `<style>`

The `<style>` element is used to specify **embedded CSS**, which we will learn about in Section 2.7.3. The contents of the `<style>` element is CSS code. Here is an example of a `<style>` element with its CSS code:

```
<style>
    p {
        font-style: italic;
    }
</style>
```

1.6.3.2   `<link>`

The `<link>` element refers to **external CSS**, which we will learn about in Section 2.7.4. The `<link>` element is an empty element that must reside inside the `<head>` element. Its important attributes are:

- `rel`—Should have the value `"stylesheet"`
- `href`—Specifies the location of a file containing CSS code
- `type`—Should have the value `"text/css"`

An example of a `<link>` element is shown below:

```
<link rel="stylesheet" href="style.css" type="text/css">
```

1.6.3.3   `<script>`

The `<script>` element is used to load **JavaScript** code, which we will learn about starting in Chapter 3. The `<script>` element may contain JavaScript code as text contents, much like the `<style>` element contains embedded CSS code as text contents (Section 1.6.3.1). Here is an example of a `<script>` element with embedded JavaScript code:

```
<script>
    function hello() {
        document
            .getElementById("demo")
            .innerHTML = "Hello JavaScript!";
    };
    document
        .getElementById("change_text")
        .addEventListener("click", hello);
</script>
```

Alternatively, the `<script>` element may contain an `src` attribute, which specifies the location of a file containing JavaScript code. This is similar to the way that the `<link>` element specifies the location of an external file with CSS code (Section 1.6.3.2). Here is an example of a `<script>` element that links to an external file with JavaScript code:

```
<script src="jquery.js"></script>
```

We will elaborate on specifying file paths in `<link>` and `<script>` elements in Section 5.5.

1.6.4   Headings and paragraphs

1.6.4.1   `<h1>`, `<h2>`, `<h3>`, `<h4>`, `<h5>`, `<h6>`

The `<h1>`, `<h2>`, ..., `<h6>` elements are block-level elements that denote that the contents are a section **heading**. The `<h1>` element is used to specify the highest, top-level headings. The `<h2>` element is used to specify second-level headings, and so on. For example, the following HTML element defines a second-level heading:

```
<h2>This is a level-2 heading!</h2>
```

1.6.4.2   `<p>`

The `<p>` element is a block-level element defining a **paragraph**. Note that the browser automatically decides where to break lines inside the paragraph, according to the containing element width (see Section 1.6.6 below). For example, the following code defines a paragraph:

```
<p>This is a paragraph!</p>
```

- Open the web page of the online version of this book, or any other web page that has lots of text.
- Resize browser window width.
- You should see the paragraph length changing, as the text is being split to multiple lines in different ways, depending on page width.

1.6.5   Font formatting

1.6.5.1   Overview

Some characteristics of font formatting can be modified using HTML elements, such as `<b>` for bold font (Section 1.6.5.2), `<i>` for italics (Section 1.6.5.3), `<sup>` for superscript (Section 1.6.5.4), `<sub>` for subscript (Section 1.6.5.5), and `<pre>` for preformatted text (Section 1.6.5.6). These characteristics, and other ones such as text color, can also be specified using CSS, which we learn later on (Sections 2.8.2–2.8.3).

1.6.5.2   `<b>`

Text within the `<b>` element appears **bold**. For example, the following HTML code:

```
This text <b>is bold</b>.
```

renders the "is bold" part in bold font.

1.6.5.3   `<i>`

Text within the `<i>` element appears **italic**. For example, the following HTML code:

```
This text <i>is italic</i>.
```

renders the "is italic" part in italic font.

1.6.5.4   `<sup>`

Text within the `<sup>` element appear superscript. For example, the following HTML code:

```
E=MC<sup>2</sup>
```

appears in the browser as:

$E{=}MC^2$

1.6.5.5 `<sub>`

Text within the `<sub>` element appear subscript. For example, the following HTML code:

```
The concentration of CO<sub>2</sub> is increasing.
```

appears in the browser as:

The concentration of CO_2 is increasing.

1.6.5.6 `<pre>`

The `<pre>` element—**preformatted** text—is a block-level element that displays any text contents exactly as it appears in the source code. This is contrary to the usual behavior, where the browser ignores line breaks and repeated spaces (Section 1.6.6). The `<pre>` element is useful for displaying computer code or computer output. For example, consider the following `<pre>` element:

```
<pre>
Text in a pre element
is displayed in a fixed-width
font, and it preserves
both      spaces and
line breaks
</pre>
```

The text is displayed in the browser as is, with the given spaces and line breaks:

```
Text in a pre element
is displayed in a fixed-width
font, and it preserves
both      spaces and
line breaks
```

Note that the `<pre>` element contents are displayed in a **fixed-width** font by default, which is convenient for computer code but may not be suitable for ordinary text.

- Try pasting the above text inside a `<p>` element, rather than a `<pre>` element, to observe the way that line breaks and multiple spaces are ignored when the text is displayed in the browser.

1.6.6 Spacing

1.6.6.1 Whitespace collapsing

When the browser comes across two or more spaces next to each other, it only displays one space. Similarly, if it comes across a line break, it treats it as a single space too. This is known as **whitespace collapsing**. Consider the following HTML code of `example-01-02.html`:

```
<!DOCTYPE html>
<html>
    <head>
        <title>White space collapsing</title>
    </head>
    <body>
        <p>The Moon's distance to Earth is 384,402 km</p>
        <p>The Moon's     distance to Earth is 384,402 km</p>
        <p>The Moon's distance to Earth

        is 384,402 km</p>
    </body>
</html>
```

All three paragraphs appear the same in a web browser (Figure 1.5) because multiple spaces and new line breaks are ignored.

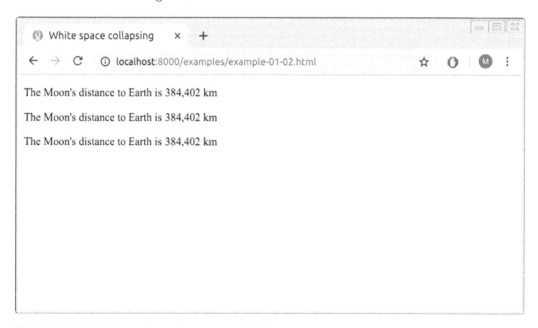

FIGURE 1.5: Screenshot of `example-01-02.html`

What if we still need to break our text, outside of a `<pre>` element? The `<br>` and `<hr>` elements can be used to do that, as shown in the next two Sections 1.6.6.2–1.6.6.3.

1.6.6.2   `<br>`

The `<br>` element is an empty element that forces a new line or **line break**. For example, the following `<br>` element will split the sentence in two lines:

`This is a new<br>line.`

The first line will contain the "This is a new" part, while the second line will contain the "line" part. The **
** element should be used sparingly; in most cases, text should be broken into lines by the browser to fit the available space.

1.6.6.3 <hr>

The **<hr>** element is an empty element that produces a **horizontal rule** (line). For example:

```
<hr>
```

A horizontal line will appear in the browser at the location where the **<hr>** element appears.

1.6.7 Lists

1.6.7.1 <ul>, <ol>, and <li>

An **unordered list** (i.e., a bullet-point list) can be created with the **<ul>** element. The **<ul>** element contains internal **<li>** elements, representing the individual *list items*. For example, the following HTML code creates an unordered list with three items:

```
<ul>
  <li>Coffee</li>
  <li>Tea</li>
  <li>Milk</li>
</ul>
```

An **ordered list** (numbered list) list can be created exactly the same way, just replacing the **<ul>** element with the **<ol>** element. For example, the following HTML code creates an ordered list with the same three items:

```
<ol>
  <li>Coffee</li>
  <li>Tea</li>
  <li>Milk</li>
</ol>
```

Figure 1.6 shows **example-01-03.html**, a web page with two headings and the two lists shown in last two code sections.

It is important to note that list items (**<li>** elements) can contain anything, not just text. For example, you can make a list of tables, a list of images, a list of lists, and so on. Lists can also be styled (using CSS) in different ways, to serve different purposes. Navigation bars and tables of contents you usually see on web pages are commonly just styled lists. For example, the sidebar with the table of contents in the online version of this book (Section 0.7) is actually a set of nested lists.

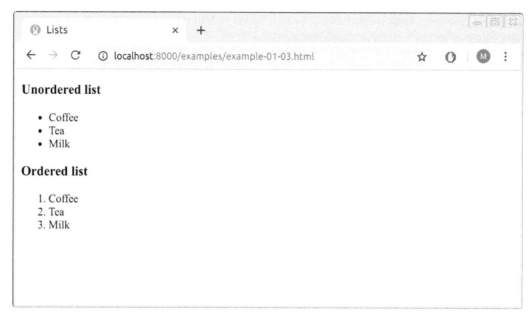

FIGURE 1.6: Screenshot of `example-01-03.html`

1.6.8   Links

1.6.8.1   `<a>`

The `<a>` element creates a **link**. Links are the defining feature of the web, because they allow you to move from one page to another—enabling the very idea of browsing or surfing. There are several types of links:

- Links from one website to another **website**
- Links from one page to another **page** on the same website
- Links from one part of a web page to another **part** of the same page
- Other types of links, such as those that start up your email program and compose a new **email** to someone

Additionally, we can distinguish between links that open in the *same* browser window and links that open in a *new* browser window.

Users can click on anything between the opening `<a>` tag and the closing `</a>` tag of a link. This means a link can be composed of text but also other clickable elements such as images. The address of the page that the link leads to is specified using the `href` (Hypertext reference) attribute. The value of an `href` attribute can be:

- An **absolute** URL which points to another website, such as `href="http://www.bgu.ac.il"`.
- A **relative** URL which points to a file within a website. For example, `href="index.html"` points to the file named `index.html` on the currently viewed website (more on URLs and website file structure in Section 5.5).
- A link to **an element** with a specified `id` (see Section 1.7.2 below) within the current web page, in which case the browser will scroll to that location. For example, `href="#top"` points to the element that has `id` value of `"top"`.

- A **combination** of a URL and location within the page. For example, `href="index.html#top"` points to the element that has `id` value of `"top"` within the file named `index.html`.

By default, the link is opened in the same browser window. If you want a link to open in a *new window*, you can add the `target="_blank"` attribute in the opening `<a>` tag.

For example, the following HTML code displays the word "BGU" as a link—usually in blue and underlined font. Clicking on the word "BGU" navigates to `http://www.bgu.ac.il` in a new browser window.

```html
<a href="http://www.bgu.ac.il" target="_blank">BGU</a>
```

- Edit the HTML file of either one of the examples shown so far (`example-01.html`, `example-02.html` or `example-03.html`) by inserting the above `<a>` element into the HTML `<body>`.
- Refresh the page.
- You should now see a functional link to the `http://www.bgu.ac.il` page.

1.6.9 Images

1.6.9.1 `<img>`

The `<img>` element adds an **image** to the web page. This is an empty, inline element, which means that images are treated like words in a sentence. The most important attribute of the `<img>` element is `src`, which specifies the file path of the image. The `src` may be a path to a local file (more on that in Section 5.5) or a URL, i.e., an image located anywhere on the web. We can also set image height and/or width using the `height` and `width` attributes, respectively.

For example, the following HTML document (`example-01-04.html`) contains one text paragraph and one image:

```html
<!DOCTYPE html>
<html>
    <head>
        <title>Images</title>
    </head>
    <body>
        <p>This is the logo of the Leaflet JavaScript library</p>
        <img src="images/leaflet.png" width="300px">
    </body>
</html>
```

The result is shown in Figure 1.7. Note that for this code to work, a local **Portable Network Graphics (PNG)** image named `leaflet.png` needs to exist. Moreover, the image placement needs to correspond to the specified file path `images/leaflet.png`, which means that the `leaflet.png` file is in the `images` folder, inside the same directory as the `index.html` file. Again, don't worry if this is not clear: we will learn about specifying file paths in Section 5.5.

FIGURE 1.7: Screenshot of `example-01-04.html`

- Edit the previous example by replacing the file `src` file path with the URL `https://leafletjs.com/docs/images/logo.png`.
- Refresh the page.
- The image should now be loaded from a remote location—the `https://leafletjs.com` website—instead of a local file.

1.6.10   Tables

1.6.10.1   `<table>`, `<th>`, `<tr>`, and `<td>`

A **table** is defined using the `<table>` element, which contains one or more `<tr>` (table **row**) elements, each of which contains one or more `<td>` (table **data**) elements. The `<td>` element contains the contents of a single table **cell**. The first table row can be defined with `<th>` (table **heading**) elements instead of `<td>` elements. The heading is shown in bold font by default. Unless explicit dimensions are given, the table rows and columns are automatically sized to fit their contents.

The following HTML code creates a table with three rows and three columns. This particular table lists three of the JavaScript libraries we are going to use later on in the book:

```
<table>
    <tr>
        <th>Library</th>
        <th>Version</th>
        <th>Released</th>
    </tr>
    <tr>
        <td>jQuery</td>
        <td>3.3.1</td>
        <td>2018-01-20</td>
    </tr>
    <tr>
        <td>Leaflet</td>
        <td>1.3.3</td>
        <td>2018-07-18</td>
    </tr>
    <tr>
        <td>Turf.js</td>
        <td>5.1.6</td>
        <td>2017-12-10</td>
    </tr>
</table>
```

The table, embedded in `example-01-05.html`, appears in the browser as shown in Figure 1.8. It may not look impressive, but keep in mind that table styling, such as alignment, border size, and color, etc., can be customized using CSS, which we learn about in Chapter 2. There are also several other element types that can go into the `<table>` element to make more complex tables[17], such as the `<caption>` element for adding a table caption.

1.6.11 Grouping

1.6.11.1 Overview

The next HTML elements we discuss are `<div>` (Section 1.6.11.2) and `<span>` (Section 1.6.11.3). These are generic block-level and inline elements, respectively, which are used to **group** other elements and content, usually to associate each group with CSS styling rules (Chapter 2), or (mostly for `<div>`) with JavaScript code (Chapter 4).

1.6.11.2 `<div>`

The `<div>` element allows you to group a set of elements together in one block-level box. For example, you might create a `<div>` element to contain all of the elements for the header of your site (the logo and the navigation bar), or you might create a `<div>` element to contain comments from visitors.

---

[17]https://developer.mozilla.org/en-US/docs/Web/HTML/Element/table

FIGURE 1.8: Screenshot of `example-01-05.html`

In a browser, the contents of the `<div>` element will start on a new line, but other than this it will make no difference to the presentation of the page. Using an `id` or `class` attribute on the `<div>` element (Section 1.7 below), however, we can distinguish the `<div>` with specific appearance and behavior. For example, we can create styling rules (with CSS) to indicate how much space the `<div>` element should occupy on the screen (Section 2.8.4.2), where it should be placed (Section 2.8.4.6), change the appearance of all the elements contained within it, and so on.

Another use case of the `<div>` element is to create an empty container, or **placeholder**, to be populated with content on page load using JavaScript. We will use this technique throughout Chapters 6–13, when creating web maps with the Leaflet JavaScript library (Section 6.5.4).

1.6.11.3   `<span>`

The `<span>` element acts like an inline equivalent of the `<div>` element. It is used to do one of the following:

- Contain a section of text where there is no other suitable element to differentiate it from its surrounding text
- Contain a number of inline elements

Again, the most common reason why people use `<span>` elements is so that they can control the appearance of the content of these elements, using CSS.

1.6.12 Input elements

1.6.12.1 Overview

HTML supports several types of **input** elements. Input elements are used to collect information from the user and thus make the web page *interactive*. Buttons, check boxes, sliders, and text inputs are all examples of input elements. The search box on `https://www.google.com` is perhaps the most well-known example of a text-input element.

Input elements can be added with the `<input>` tag. The `<input>` element has several important attributes:

- `type`—The **type** of input
- `name`—The **identifier** that is sent to the server when you submit a form—a collection of related input elements (see below)
- `value`—The **initial** value in text and numeric inputs, or the **text** appearing on a button

Text area inputs are a special case, defined with the `<textarea>` element rather than with the `<input>` element (Section 1.6.12.5). Dropdown menu input is another special case, defined with the `<select>` and `<option>` elements (Section 1.6.12.8).

Input elements are commonly grouped inside a **form**, using the `<form>` element. This has several advantages[18] for handling multiple inputs as a single unit. Through most of the book we will use simple, individual inputs—therefore to simplify the material we will avoid enclosing the inputs in a `<form>`. We will then come back to an example with a `<form>` element in Section 13.5.

Common input types are summarized in Table 1.5 and described in more detail in Sections 1.6.12.2–1.6.12.9. Note that there are many other possible input types[19] that we will not use in this book, including specialized input elements for picking colors, selecting dates, etc.

TABLE 1.5: HTML input elements

Input type	Usage
Numeric input	`<input type="number">`
Range input	`<input type="range">`
Text input	`<input type="text">`
Text area	`<textarea></textarea>`
Radio buttons	`<input type="radio">`
Checkboxes	`<input type="checkbox">`
Dropdown lists	`<select><option></option></select>`
Buttons	`<input type="button">`

1.6.12.2 Numeric input

A numeric `<input>` element is used to get **numeric** input through typing or clicking the up/down buttons. A numeric input is defined using an `<input>` element with a `type="number"` attribute. Other important attributes are `min` and `max`, specifying the

---

[18]`https://stackoverflow.com/questions/31066693/what-is-the-purpose-of-the-html-form-tag`
[19]`https://developer.mozilla.org/en-US/docs/Web/HTML/Element/input`

valid range of numbers that the user can enter. For example, the following HTML code creates a numeric input, where the user can enter numbers between 0 and 100, with the initial value set to 5:

```
<input type="number" name="num" value="5" min="0" max="100">
```

The `name` attribute identifies the form control and is sent along with the entered information when submitting a form to a *server*. It is not very useful within the scope of this book but is shown here for completeness as it is commonly used in other contexts (Section 1.6.12.1).

The way that the above numeric input element appears in the browser, along with all other types of input we cover next (Sections 1.6.12.3–1.6.12.9), is shown in Figure 1.9. The numeric input is in the top-left corner if the figure. Note that the code for `example-01-06.html` includes CSS styling rules (which we learn about in Chapter 2) for arranging the input elements in three columns.

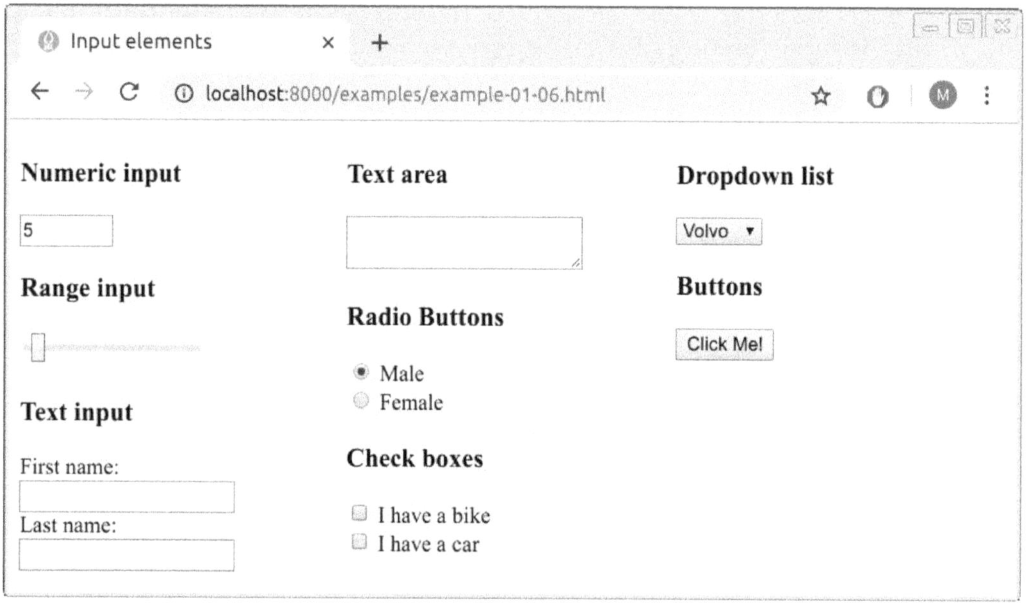

FIGURE 1.9: Screenshot of `example-01-06.html`

1.6.12.3 Range input

A range `<input>` element is used for picking numeric values with a **slider**. This is usually more convenient and intuitive for the user in cases when the exact value is not important. A range input is defined using `type="range"`. The purpose of the `value`, `min`, and `max` attributes is to specify the initial, minimal, and maximal values, respectively, just like in the numeric input (Section 1.6.12.2). Here is an example of a range input element:

```
<input type="range" name="points" value="5" min="0" max="100">
```

The result is shown in Figure 1.9.

1.6.12.4 Text input

A text `<input>` is used for typing plain **text**. A text input is defined using `type="text"`. For example, the following HTML code creates two text input boxes for entering first and last names, along with the corresponding labels[20]. The `<br>` element is used to place each text input box on a new line, beneath its label:

```
First name:<br>
<input type="text" name="firstname"><br>
Last name:<br>
<input type="text" name="lastname">
```

The result is shown in Figure 1.9.

1.6.12.5 Text area

A **text area** input is used for typing plain text, just like text input, but intended for multi-line rather than single-line text input (e.g., Figure 7.5). A text input is defined using the `<textarea>` element, as shown in the following example:

```
<textarea name="mytext"></textarea>
```

The result is shown in Figure 1.9.

1.6.12.6 Radio buttons

Radio buttons are used to select *one* of several options. Each radio button is defined with a separate `<input>` element using `type="radio"`. The user can select only one option of the radio buttons sharing the same value for the **name** attribute. The **checked** attribute can be used to define which button is selected on page load. Note that the **checked** attribute has no value. For example, the following HTML code creates two radio buttons, with corresponding labels:

```
<input type="radio" name="gender" value="male" checked> Male<br>
<input type="radio" name="gender" value="female"> Female<br>
```

The result is shown in Figure 1.9. The "Male" option is initially checked because of the **checked** attribute.

---

[20]Labels for input elements can also be created using the specialized `<label>` element (`https://developer.mozilla.org/en-US/docs/Web/HTML/Element/label`), rather than using simple text as shown in the example. The advantage of `<label>` is that clicking on the text within the label triggers the associated input. This is not crucial for the purposes of this book, so we will use plain text labels rather than `<label>` elements for simplicity, as shown in the example.

1.6.12.7   Checkboxes

Checkboxes are used to select *one or more* (or none) of several options. Each checkbox is defined with a separate `<input>` element using `type="checkbox"`. For example, the following HTML code creates two checkboxes, with labels:

```
<input type="checkbox" name="vehicle1" value="Bike"> I have a bike<br>
<input type="checkbox" name="vehicle2" value="Car"> I have a car<br>
```

The result is shown in Figure 1.9.

1.6.12.8   Dropdown menus

Dropdown lists, or dropdown menus, are used to select *one* option from a *list*. The list is initially hidden from view, expanding only when clicked. The list is also scrollable, therefore the number of items is potentially longer than can fit on screen. This makes dropdown lists suitable for situations when we have a long list of options the user needs to choose from, and we do not want to "waste" page space displaying all possible options at all times (e.g., Figure 10.4).

The dropdown menu is initiated using the `<select>` element. Inside the `<select>` element, each option is defined with a separate `<option>` element. For example:

```
<select name="cars">
    <option value="volvo">Volvo</option>
    <option value="suzuki">Suzuki</option>
    <option value="fiat">Fiat</option>
    <option value="audi">Audi</option>
</select>
```

The result is shown in Figure 1.9.

Note that in radio buttons (Section 1.6.12.6), checkboxes (Section 1.6.12.7) and dropdown menus (Section 1.6.12.8), the `value` attribute identifies the currently selected option when sending the data to the server. The `value` does not necessarily have to be identical to the text contents we see on screen when interacting with the input element in the browser. For example, in the above HTML code the first `<option>` has `value="volvo"`, which is used to identify the option when sending data to a server, while the text shown on screen is actually `"Volvo"` (with capital V).

1.6.12.9   Buttons

A **button** is used to trigger actions on the page. A button can be created using the `<input>` element with the `type="button"` attribute. The `value` attribute is used to set the text label that appears on the button. For example, the following HTML code creates a button with the text "Click me!" on top:

```
<input type="button" value="Click Me!">
```

The result is shown in Figure 1.9.

On their own, the input elements are not very useful. For example, interacting with the various input elements in `example-01-06.html` (Figure 1.9) has no effect whatsoever. To make the input elements useful, we need to capture the input element values and write code that does something with those values. In Section 4.14 we will learn how the current values of input elements can be captured and used to modify page appearance and/or contents, using JavaScript.

1.7 id, class, and style attributes

1.7.1 Overview

So far we have mostly encountered *specific* attributes for different HTML elements. For example, the `src` attribute is specific to `<img>` (and several other) elements and the `href` attribute is specific to `<a>` (and several other) elements. All HTML elements also share three important *non-specific* attributes, which can appear in any element:

- `id`—Unique identifier
- `class`—Non-unique identifier
- `style`—Inline CSS

The following Sections 1.7.2–1.7.4 cover the purpose and usage of these three non-specific attributes.

1.7.2 id

The `id` attribute is used to *uniquely* identify an HTML element from other elements on the page. Its value should start with a letter or an underscore, not a number or any other character. It is important that no two elements on the same page have the same value for their `id` attributes—otherwise the value is no longer unique.

For example, the following page has three `<p>` elements with `id` attributes. Note that the values of the `id` attribute—`"intro"`, `"middle"`, and `"summary"`—are different from each other and thus unique for each element.

```html
<!DOCTYPE html>
<html>
    <head>
        <title>A Minimal HTML Document</title>
    </head>
    <body>
        <p id="intro">The 1st paragraph is an overview.</p>
        <p id="middle">The 2nd paragraph gives more details.</p>
        <p id="summary">The 3rd paragraph is a summary.</p>
    </body>
</html>
```

As we will see when discussing CSS (Chapter 2), giving an element a unique `id` allows us to *style* it differently than any other instance of the same element on the page. For example, we may want to assign one paragraph within the page a different color than all of the other paragraphs. When we go on to learn about JavaScript and interactive behavior (Chapter 4), we will also use `id` attributes to allow our scripts to uniquely affect the *interactive behavior* of particular elements on the page.

1.7.3   class

Every HTML element can also carry a `class` attribute. Sometimes, rather than uniquely identify one element within a document using an `id`, we will want to identify a *group* of elements as being different from all other elements on the page. For example, we may have some paragraphs of text that contain information that is more important than others and want to distinguish these elements, or differentiate between links that point to other pages on your own site and links that point to external sites.

To mark multiple elements as belonging to one group we can use the `class` attribute. The value of the `class` attribute identifies the group those elements belong to. For example, in the following HTML document, the first and third `<p>` elements share the `class` attribute value of `"important"`.

```
<!DOCTYPE html>
<html>
    <head>
        <title>A Minimal HTML Document</title>
    </head>
    <body>
        <p class="important">The 1st paragraph is an overview.</p>
        <p>The 2nd paragraph gives more details.</p>
        <p class="important">The 3rd paragraph is a summary.</p>
    </body>
</html>
```

Just like an `id`, the `class` attribute is commonly used for styling, or interacting with, a group of elements on the page.

1.7.4   style

All elements may also have a `style` attribute, which allows **inline CSS** rules to be specified within the element's start tag. We will talk about inline CSS in Section 2.7.2.

1.8   Code layout

When writing code, it is useful to keep a uniform code layout. For example, we can use tabs to distinguish content that is inside another element, thus highlighting the hierarchical structure of code.

The following two HTML documents are the same as far as the computer is concerned, i.e., they are displayed exactly the same way in the browser. However, the second HTML document is much more readable to humans thanks to the facts that:

- Each element starts on a new line.
- Internal elements are indented with tabs.

```
<!DOCTYPE html><html><head><title>A Minimal HTML Document</title></head>
<body><p>The content goes here!</p></body></html>
```

```
<!DOCTYPE html>
<html>
    <head>
        <title>A Minimal HTML Document</title>
    </head>
    <body>
        <p>The content goes here!</p>
    </body>
</html>
```

1.9   Inspecting elements

When looking at the HTML code of a simple web page, such as the ones we created in this chapter, it is easy to locate the HTML element responsible for creating a given visual element we see on screen. However, as the HTML code becomes longer and more complex, it may be more difficult to make this association.

Luckily, browsers have a built-in feature for locating HTML code associated with any element you see on screen. For example:

- Open the example file named `example-01-01.html` in Chrome.
- Press **Ctrl+Shift+I** or **F12**.

The screen should now be split. The left pane still shows the web page. The right pane shows the **developer tools**. The developer tools are a set of web authoring and debugging tools built into modern web browsers, including Chrome. The developer tools provide web developers access into the internals of the browser and the web page being displayed.

- Press **Ctrl+Shift+C**.

This toggles the **Inspect Element** mode. (It also opens the developer tools in the Inspect Element mode if they are not already open.) In the Inspect Element mode, you can *hover* above different parts of the page (left pane) with the mouse pointer. The relevant elements are highlighted, and their name is shown (Figure 1.10). Clicking on an element highlights the relevant part of the page source code and scrolls it into view. This also works in the opposite direction: hovering over the code in the right pane highlights the respective visual element in the left pane.

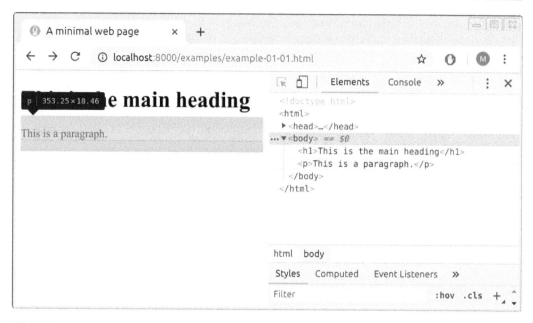

FIGURE 1.10: Using the *Inspect Element* tool in Chrome

Remember how we mentioned that every HTML element can be thought of as a horizontal box, where (by default) *height* is determined by amount of content and *width* is set to maximum of browser width (Section 1.5.3)? This becomes evident when the Inspect Element tool highlights those boxes (Figure 1.10).

1.10   Exercise

- Edit the minimal HTML document `example-01-01.html` to experiment with the HTML element types we learned in this chapter:
    - Modify the title of the page and the first-level heading.
    - Delete the existing paragraph and add a new paragraph with two to three sentences about a subject you are interested in.
    - Use the appropriate tags to format some of the words in *italic* or **bold** font.
    - Use the `<a>` tag to add a link to another web page.
    - Add a list with two levels, i.e., a list where each list item is also a list.
    - Add images which are loaded from another location on the internet, such as from **Flickr**[21].

---

[21]https://www.flickr.com/

2

CSS

2.1 Introduction

In Chapter 1, we introduced HTML—the data format for specifying the contents and structure of web pages. Using HTML alone, however, the web pages you create will share the same default single-column arrangement, default colors, default fonts, etc. You will almost never encounter "real" websites with the outdated appearance like the examples shown in Chapter 1. The main thing that differentiates modern-looking, customized web pages from simple ones is **CSS**—the data format for specifying style and appearance of web pages.

2.2 What is CSS?

Cascading Style Sheets (CSS)[1] is a language for specifying the *style* and *appearance* of web pages. For example, CSS is used to specify:

- Fonts
- Colors
- Arrangement of elements on the page

The key to understanding how CSS works is to imagine that there is an *invisible box* around every HTML element (Figure 2.1). CSS allows us to create rules that control the way that each individual box, and the contents of that box, is presented. CSS is run when it is linked to HTML code, while that HTML code is processed and visually presented by the browser.

2.3 CSS rules

CSS works by associating styling **rules** with HTML elements. The CSS rules govern how the contents of specified elements should be displayed. A CSS rule contains two parts:

- A **selector**
- One or more **declarations**

---

[1] https://en.wikipedia.org/wiki/Cascading_Style_Sheets

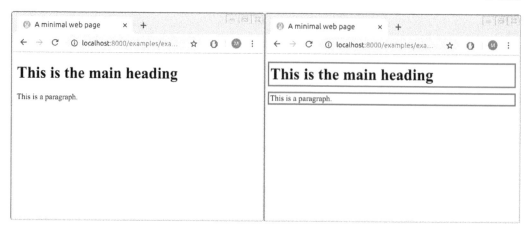

FIGURE 2.1: HTML elements considered as boxes

The selector specifies which HTML elements the rule applies to. The declarations indicate how the elements referred to by the selector should be styled. Each declaration is split into two parts: a property *name* and a property *value*. Here is an example of a CSS rule:

```
a { color: white; }
```

This particular rule indicates that all `<a>` elements, i.e., links (Section 1.6.8.1), will be displayed with white color. Let's go over the structure of this rule:

- a is the **selector**, specifying that the rule applies to all `<a>` elements.
- Curly brackets `{...}` contain the **declarations**.
- In this case there is only one declaration, `color: white;`, which means that the affected elements will be colored in white. The declaration is composed of:
  - A property **name** (`color`) indicating the property that we want to change
  - A property **value** (`white`) specifying the settings we want to use for the chosen property
  - The property name and property value are separated with :
  - The declaration ends with ;

Here is another example:

```
a { font-style: italic; }
```

This time the rule indicates that all `<a>` elements should be shown in italics. The only difference from the previous example is that the declaration `color: white;` was replaced with the declaration `font-style: italic;`. Note the different components of the CSS rule, most importantly the selector, the property name, and the property value (Table 2.1).

TABLE 2.1: CSS rule structure

Component	Example
CSS rule	a { font-style: italic; }
Selector	<u>a</u> { font-style: italic; }
Declaration	a { <u>font-style: italic;</u> }

TABLE 2.1: CSS rule structure

Component	Example
Property name	a { <u>font-style</u>: italic; }
Property value	a { font-style: <u>italic</u>; }

2.4 CSS selectors

2.4.1 Overview

Within a CSS rule, the selector specifies which HTML elements will be affected by the rule. There are many different types of CSS selectors that allow you to target rules to a very specific selection of elements in an HTML document. Some CSS selectors are used to target specific HTML elements (Table 2.2), while other selectors specify selector *combinations* (Table 2.3).

TABLE 2.2: CSS selectors targeting specific elements

Selector	Meaning	Example
Universal Selector	Applies to all elements in the document	`*` `{}` targets all elements on the page
Type Selector	Matches element names	`h1 {}` targets all `<h1>` elements
Class Selector	Matches all elements whose `class` matches the one specified after the dot (.)	`.note {}` targets any element whose `class` attribute has a value of "note". `p.note {}` targets only `<p>` elements whose `class` attribute has a value of "note"
ID Selector	Matches an element whose id matches the one specified after the hash (#)	`#introduction {}` targets the element whose `id` attribute has the value of "introduction"

TABLE 2.3: CSS selector combinations

Selector	Meaning	Example
Multiple Selector	Matches elements corresponding to any of the selectors	`h1, h2, h3 {}` targets the `<h1>`, `<h2>`, and `<h3>` elements
Descendant Selector	Matches an element that is a descendant of another element (not necessarily direct child)	`p a {}` targets all `<a>` elements that sit inside a `<p>` element, even if there are other elements between them
Child Selector	Matches an element that is a direct child of another element	`li>a {}` targets all `<a>` elements that are children of `<li>` elements

Selector	Meaning	Example
Adjacent Sibling Selector	Matches an element that is the nearest sibling of another	`h1+p {}` targets those `<p>` elements which come right after an `<h1>` element
General Sibling Selector	Matches an element that is a sibling of another	`h1~p {}` targets all `<p>` elements that follow, immediately or not, an `<h1>` element

Throughout this book we will use the following most commonly used selectors:

- Type selector, possibly combined in a multiple selector (`h1, h2, h3 {}`)
- Class selector (`.note {}`)
- ID selector (`#introduction {}`)
- Descendant selector (`p a {}`)

More details on these selectors are given in the following Sections 2.4.2–2.4.5.

2.4.2   Type selector

The **type selector** applies to the element(s) of the specified type in the linked HTML code. In case there are more than one, the type names should be separated by a comma, in a combination known as a **multiple selector** (Table 2.3). For example, the following rule applies to all `<a>` elements:

```
a { color: white; }
```

The following rule, however, applies to all `<a>` *and* all `<p>` elements:

```
p, a { color: white; }
```

2.4.3   Class selector

The **class selector** contains a full stop (.), and the part after the full stop specifies the name of a class (Section 1.7.3). All elements that have a `class` attribute with the given value will be affected by the rule. For example, the following rule applies to all elements that have the attribute `class="figure"`:

```
.figure {
    display: block;
    margin-left: auto;
    margin-right: auto;
}
```

We can combine the class selector with the type selector. This means the rule will no longer apply to all elements with the appropriate class, but only to the HTML elements of a particular type *and* the appropriate class. For example, the following rule applies only to the `<img>` elements that have `class="figure"`:

```
img.figure {
    display: block;
    margin-left: auto;
    margin-right: auto;
}
```

2.4.4   ID selector

The **ID selector** contains a hash character (#) followed by an ID. The rule applies to the specific element that has the given `id` attribute. The ID selector can be used to control the appearance of exactly *one* element, since the `id` is unique (Section 1.7.2). For example, the following rule applies to the specific element which has the attribute `id="footer"`:

```
#footer { font-style: italic; }
```

2.4.5   Descendant selector

The **descendant selector**[2] contains several element names, separated by spaces. This allows a CSS rule to be applied to an element only when the element is contained *inside* another type of element. For example, the following rule applies to all `<a>` elements that are within `<p>` elements:

```
p a { color: white; }
```

Note that, in the above example, the `<a>` element does not need to be immediately within the `<p>` element, unlike when using the **child selector** (Table 2.3).

2.5   CSS conflicts

It is possible for CSS rules to **conflict**, i.e., for there to be more than one rule affecting the same *property* in the same *element*. In case of conflict, generally a *more specific* rule will override a less specific one. For example, an ID selector will override a type selector, since the former is more specific. Given the following pair of CSS rules, a `<p>` element with `id="intro"` will be colored red according to the more specific rule, while all other `<p>` elements will be colored black:

---

[2]The term *descendant* refers to an HTML element that is contained within another given HTML element, i.e., "downstream" in the HTML document hierarchy.

```
p { color: black; }
#intro { color: red; }
```

If there are two rules having the same level of specificity, the rule that is specified *last* wins. For example, if two CSS files are linked in the header of an HTML document (see Section 2.7.4 below) and they both contain rules with the same selector, then the rule in the second file—the one processed later by the browser—will override the rule in the first file. As another example, consider the pair of CSS rules shown below. In this case, all `<p>` element will be colored red—according to the *last* specified rule:

```
p { color: black; }
p { color: red; }
```

The next example is more complex, containing eight CSS rules with numerous conflicts:

```
* { font-family: Arial, Verdana, sans-serif; }
h1 { font-family: "Courier New", monospace; }
i { color: green; }
i { color: red; }
b { color: pink; }
p b { color: blue; }
p#intro { font-size: 100%; }
p { font-size: 75%; }
```

Ignore the properties for now, only consider the selectors. Based on the conflict-resolving guidelines described previously, here is how this particular set of rules will be processed by the browser:

- The **h1** rule will override the ***** rule, since **h1** is more specific than *****.
  - Therefore, first-level headings will be shown in `Courier New` font rather than `Arial`.
- The **p b** rule will override the **b** rule, since **p b** is more specific than **b**.
  - Therefore, bold text inside a paragraph will be shown in blue rather than pink.
- The **p#intro** will override the **p** rule, since **p#intro** is more specific than **p**.
  - Therefore, the paragraph with `id="intro"` will be shown with larger font.
- The second **i** selector will override the first one, since it is specified later.
  - Therefore, *italic* text will be shown in red rather than green.

An example of a web page where the above CSS rules are applied (`example-02-01.html`) is shown in Figure 2.2.

2.6   CSS inheritance

Some CSS property values are **inherited**, since it makes sense to specify them for an element as well as all of the element's descendants. Other CSS properties are *not* inherited, since it makes sense to specify them just for the targeted element and not for its descendants.

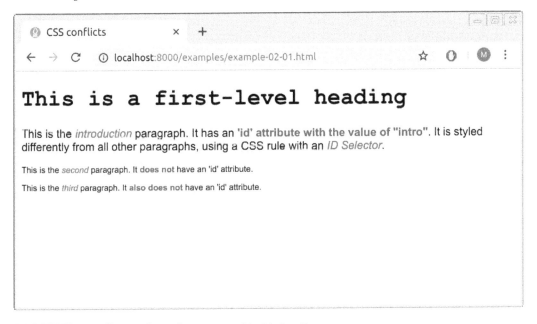

FIGURE 2.2: Screenshot of `example-02-01.html`

For example, it makes sense for `font-family` and `color` to be inherited, as that makes it easy to set a document-wide *base font* and *base color* by setting the `font-family` and `color` on the `<body>` element. That way, all of the heading, paragraphs, and all other elements in the document—which are all descendants of the `<body>` element—are uniformly displayed in the same font and color. When necessary, you can override the fonts and/or colors on individual elements where needed, relying on the fact that a specific rule overrides a general one (Section 2.5). It would be really annoying to have to set the base font separately on every element.

Contrariwise, it makes no sense for `margin`, `padding`, and `border` to be inherited. These CSS properties control the placement and border style of HTML elements (Section 2.8.4). Imagine the styling mess that would occur if you set these properties on a container element and had them inherited by every single internal element! Nevertheless, in rare situations forcing inheritance is necessary. Forcing a property to inherit values from their "parent" element can be done by using the keyword `inherit` as the respective property value (see Section 2.9 for an example)[3].

2.7 Linking CSS to HTML

2.7.1 Methods for linking CSS

How can we actually **link** our CSS code to an HTML document, to affect its presentation in the browser? There are three methods in which CSS code can be linked to an HTML document:

---

[3]A more detailed description of CSS conflict resolution and inheritance can be found in the *Cascade and Inheritance* article (`https://developer.mozilla.org/en-US/docs/Learn/CSS/Introduction_to_CSS/Cascade_and_inheritance`) by Mozilla.

- **Inline** CSS (Section 2.7.2)
- **Embedded** CSS (Section 2.7.3)
- **External** CSS (Section 2.7.4)

The following Sections 2.7.2–2.7.4 shortly explain each of these methods.

2.7.2   Inline CSS

The simplest approach to associate CSS code with HTML elements is to include the CSS code within the `style` attribute of an HTML element. As mentioned in Section 1.7.4, `style` is one of three general attributes that can appear in every type of HTML element (along with `id` and `class`). The value of the `style` attribute contains CSS rules applying to that element. This method is called **inline CSS**. For example, a paragraph that has the following start tag will be displayed with italic font, thanks to the inline CSS rule. Here, CSS is used to control the appearance of text within *this* paragraph only.

```
<p style="font-style:italic;">
```

Note that, with inline CSS, the code consists of just declaration(s), with no selectors. Selectors are not needed, since inline CSS by definition only applies to a specific HTML element where the `style` attribute appears.

Using inline CSS is generally not recommended, and should be used sparingly, because it:

- Mixes CSS code with HTML code, making both of them harder to maintain
- Requires repeating CSS code in all elements where the same style needs to be applied

We will see one example where inline CSS comes in handy later on (Section 8.6).

2.7.3   Embedded CSS

It is also possible to include CSS code within a `<style>` element within the `<head>` element of HTML code (Section 1.6.3.1). This approach is called **embedded CSS**. For example, here are the first few lines of an HTML document that uses embedded CSS:

```
<!DOCTYPE html>
<html>
    <head>
        <style>
            p { font-style: italic; }
        </style>
...
```

In this example, the appearance of text within *all* paragraphs in the document is controlled with a single CSS rule. Embedded CSS is a better approach than inline CSS, because with embedded CSS the rules do not need to be attached to each individual HTML element. However, embedded CSS is still not ideal, because:

- the separation between HTML and CSS is not complete—although all CSS code is in one place, it is still *embedded* in the HTML document; and

- consequently, any reuse of the CSS code with other HTML code requires making multiple copies of the same CSS code in the `<head>` of each page.

2.7.4   External CSS

The third approach, called **external CSS**, is writing CSS code in a separate `.css` file and referring to that CSS file from the HTML code. The CSS file is referred to using a `<link>` element within the `<head>` element (Section 1.6.3.2). For example, here are the first few lines of an HTML document that uses external CSS:

```
<!DOCTYPE html>
<html>
    <head>
        <link rel="stylesheet" href="style.css" type="text/css">
...
```

The above `<link>` element refers to a file with CSS code, in this case a file named `style.css`. The contents of the `style.css` file is plain CSS code, such as:

```
p { font-style: italic; }
```

The external CSS approach is the recommended way of using CSS, since it makes HTML and CSS code completely *separate*, and thus the CSS code is easier to maintain. Additionally, external CSS makes the code *reusable*, since the same CSS file can be linked to multiple HTML files. That way, for example, the same styling can be applied to all pages of a given website, while still keeping the CSS code in one place rather than repeating it. For example, the online version of this book (Section 0.7) is composed of multiple HTML documents (one for each chapter), all of them linked to the same external CSS files and thus sharing the same style.

In this book, we will mostly use embedded CSS (Section 2.7.3), since our code examples will be relatively short and since embedded CSS makes the presentation of the material easier: the entire source code, including HTML and CSS, can be viewed in one place in a single code file. The reader should keep in mind, however, that when working on larger projects keeping CSS in external files is the recommended way to go.

2.8   CSS properties

2.8.1   Overview

So far we have discussed the general structure of CSS code (Section 2.3), selector types (Section 2.4), conflict and inheritance issues (Sections 2.5–2.6), and ways to associate CSS code with an HTML document (Section 2.7). In this section, we will go over specific, commonly used CSS properties. The CSS properties we are going to meet can be divided into three groups, according to the styling aspect they influence:

- Color (Section 2.8.2)
- Text (Section 2.8.3)
- Boxes (Section 2.8.4)

Table 2.4 lists the specific CSS properties that we are going to cover. Note that there are many other CSS properties that we are not going to cover since they are less relevant to web mapping[4].

TABLE 2.4: Common CSS properties

Category	Property	Description
Color	`color:`	Foreground color
	`background-color:`	Background color
Text	`font-family:`	Font family
	`font-style:`	`normal` or `italic`
	`font-weight:`	`normal` or `bold`
	`font-size:`	Font size
	`text-align:`	Text alignment (e.g., `center`)
Boxes	`width:`	Box width
	`height:`	Box height
	`border-width:`	Border width
	`border-style:`	Border style (e.g., `dotted`)
	`border-color:`	Border color
	`border-radius:`	Border radius
	`margin:`	Margin size
	`padding:`	Padding size
	`position:`	Positioning method (e.g., `fixed`)
	`top:`	Top offset
	`bottom:`	Bottom offset
	`right:`	Right offset
	`left:`	Left offset
	`z-index:`	Z-index

2.8.2 Color

2.8.2.1 `color:`, `background-color:`

The `color:` property controls the **foreground color**. The `background-color:` property controls the **background color**. For example, when referring to a `<p>` element, `color:` specifies the color of text inside the paragraph, while `background-color:` specifies the background color of the paragraph "box" (Figure 2.3). The most common ways of specifying the color itself, i.e., the *values* of these properties, are summarized in Table 2.5 and described as follows.

---

[4]The *Introduction to CSS* (https://developer.mozilla.org/en-US/docs/Learn/CSS/Introduction_to_CSS) tutorial and the *CSS Reference* (https://developer.mozilla.org/en-US/docs/Web/CSS/Reference) by Mozilla can be referred to for more information on CSS.

TABLE 2.5: Methods for specifying color in CSS

Method	Example
RGB	`rgb(255,255,0)`
RGBA	`rgba(255,255,0,1)`
HEX	`#ffff00`
Color name	`yellow`

RGB values express colors in terms of how much red, green, and blue are used to make them up. The values can be integers (from 0 to 255) or percentages. For example: `rgb(255,255,0)` means yellow, since red and green are set to maximum, while blue is set to zero; `rgb(100%,100%,0%)` is another way of specifying the same color. **RGBA** values are similar to RGB values, but with an extra channel, known as the alpha channel, representing transparency (0 = fully transparent, 1 = fully opaque). For example: `rgba(0,0,0,0.5)` means 50% transparent black, i.e., grey.

HEX codes are six-digit codes that represent the amount of red, green, and blue in a color, preceded by a hash # sign. Each pair of digits specifies a number between 0 and 255 in hexadecimal[5] notation. For example, `#ffff00` specifies yellow color, since in hexadecimal notation `ff` specifies 255 and `00` specifies 0. RGB(A) and HEX are, therefore, identical in their ability to convey $256^3 = 16,777,216$ different colors[6]. The difference between the two methods is that RGB uses the *decimal* number notation while HEX uses the *hexadecimal* notation[7].

Finally, there are several predefined **color names** which are recognized by browsers, such as `black`, `white`, `red`, `green`, `blue`, and `yellow`. It should be noted that using predefined colors such as `red` and `blue` is usually not ideal in terms of graphic design. RGB and HEX color specifications should be generally preferred, as they make it possible to make subtle color optimization and to use predefined, carefully chosen color scales, as we will see later on (Section 8.4.3). The *X11 Color Names*[8] Wikipedia entry gives a list of predefined color names along with their RGB and HEX codes. It is useful to go over it, to get a feeling of the three color-specification systems[9].

To summarize the four color-specification methods, the following CSS code specifies the `color` property for `<h1>` and `<h2>` elements, and the `color` as well as the `background-color` of the `<p>` elements on the page. Note how the code uses all four above-mentioned methods for specifying colors with CSS:

```css
h1 {
    color: DarkCyan;  /* Color name */
}
h2 {
    color: #ee3e80;   /* HEX */
}
```

---

[5]https://en.wikipedia.org/wiki/Web_colors#Hex_triplet

[6]https://en.wikipedia.org/wiki/RGB_color_model

[7]An interactive demonstration of the HEX and RGB notations can be found in the *Colors HEX* reference (https://www.w3schools.com/colors/colors_hexadecimal.asp) by W3Schools.

[8]https://en.wikipedia.org/wiki/Web_colors#X11_color_names

[9]The *Color Reference* (https://developer.mozilla.org/en-US/docs/Web/CSS/color_value) by Mozilla gives more examples of specifying color in CSS.

```
p {
    background-color: rgba(255,255,0,0.8);   /* RGBA */
    color: rgb(100,100,90);                  /* RGB */
}
```

A small example, `example-02-02.html`, demonstrating the effect of these rules, is shown in Figure 2.3.

FIGURE 2.3: Screenshot of `example-02-02.html`

Note that we used CSS comments in the last example. CSS comments are written between the `/*` and the `*/` symbols, like so:

```
/* This is a CSS comment */
```

2.8.3   Text

2.8.3.1   font-family:

The `font-family:` property controls the **font** for text within an element. The value of the `font-family:` property can be:

- A **specific** font family name, such as `"Times New Roman"`, `Courier`, or `Arial`
- A **generic** font family name, such as `serif`, `sans-serif`, or `monospace`

It is a good idea to specify several values, separated by commas, to indicate that they are alternatives. The browser will select the first font in the list that is installed or that can be downloaded. The recommendation[10], which we follow here, is to put quotes around font names that contain spaces, such as `"Times New Roman"` instead of `Times New Roman`.

The last option on the list is usually a generic font family name. In case the person viewing the result does not have any of the specific fonts on their computer, the browser will fall back to any available font from the generic family. For example:

```
.serif { font-family: Times, "Times New Roman", Georgia, serif; }
.sansserif { font-family: Verdana, Arial, Helvetica, sans-serif; }
.monospace { font-family: "Lucida Console", Courier, monospace; }
```

In this example, the first rule means that a `Times`, `"Times New Roman"`, or `Georgia` font will be used if available (ordered from highest to lowest preference). Otherwise, the browser will choose any `serif` font that is available[11].

2.8.3.2 `font-style:`, `font-weight:`, `font-size:`

These properties control the detailed **appearance** of text. The `font-style:` can be `normal` or `italic`. The `font-weight:` can be `normal` or `bold`. The `font-size:` determines *font size*, which can be specified in several ways:

- In **pixels**, for example: 12px
- In **percentages** relatively to the default, for example: 75%
- In **ems**, which is equivalent to the width of the letter M in the current font size, for example: 1.3em

2.8.3.3 `text-align:`

The `text-align:` property controls the **alignment** of text within an element, with possible values `left`, `right`, `center`, or `justify`.

The following example includes several of the above-mentioned CSS properties related to text:

```
body {
    font-family: "Lucida Console", Courier, monospace;
}
h2 {
    font-family: "Times New Roman", Times, serif;
    font-style: italic;
}
p {
    font-size: 19.2px;
    text-align: center;
}
```

---

[10]https://www.w3.org/TR/CSS21/fonts.html#propdef-font-family

[11]More details on the `font-family:` property can be found in the *font-family* reference (https://developer.mozilla.org/en-US/docs/Web/CSS/font-family) by Mozilla.

The `example-02-03.html` web page demonstrates the visual effects of the these CSS rules (Figure 2.4).

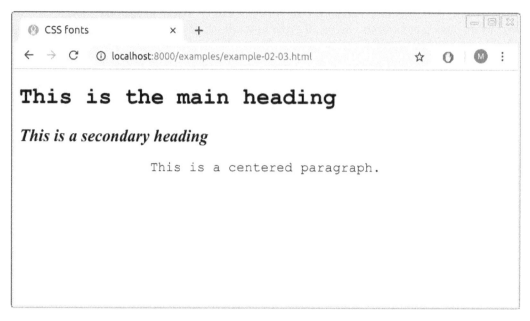

FIGURE 2.4: Screenshot of `example-02-03.html`

2.8.4   Boxes

2.8.4.1   CSS box properties

In the beginning of this chapter, we mentioned how CSS treats each HTML element as if it lives in its own box (Figure 2.1). You can set several properties that affect the appearance and placement of these boxes. For example, using CSS it is possible to:

- Control the **dimensions** of boxes
- Create **borders** around boxes
- Set **margins** and **padding** for boxes
- Place the boxes in different **positions** on the page

The terms margin, border, and padding can be confusing[12], so we will define them right from the start. Every box has three properties that can be adjusted to control its boundaries (Figure 2.5):

- **Margin**—Margins sit outside the edge of the border. You can set the width of a margin to create a gap between the borders of two adjacent boxes. Margins are always transparent.
- **Border**—Every box has a border, even if it is not visible or is specified to be 0 pixels wide. The border separates the edge of one box from another.
- **Padding**—Padding is the space between the border of a box and any content contained within it. Adding padding can increase the readability of the contents.

---

[12]For a good discussion of the CSS box model, see also the *CSS Positioning* article (http://www.brainjar.com/css/positioning/).

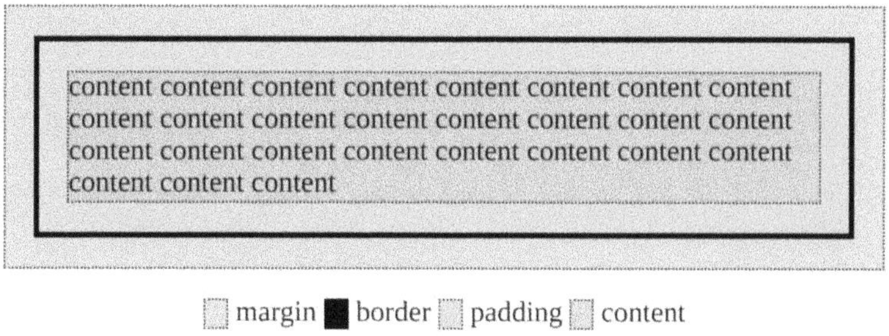

margin █ border ▢ padding ▢ content

FIGURE 2.5: Margin, border, and padding around an HTML element

2.8.4.2 `width:`, `height:`

By default, a box is sized just big enough to hold its contents. For example, a paragraph of text expands to fill the entire width of the page and uses as many lines as necessary (Section 1.6.4.2), while an image has an *intrinsic size*—the number of pixels in each direction (Section 1.6.9.1). The `width:` and `height:` properties provide means to override the default width and height and explicitly control the dimensions of an element.

Widths or heights can be specified either as *percentages* of the parent element, or as *absolute* values (such as in `px` units). For example, within a web page that is 800 pixels wide, to make a paragraph of text use half of the page width, we can use either of the following specifications:

```
p { width: 50% }
p { width: 400px }
```

It is important to note that element `width:` and `height:` determine the content size, so if we also add padding (Section 2.8.4.5) and borders (Section 2.8.4.3) then the final size of the element will be larger than intended. There is a property named `box-sizing:` which solves this problem by specifying the total size, including content, padding, and border.

The following HTML document `example-02-04.html` uses the `width:` and `height:` properties to create three `<div>` elements, each having 100px height and 20% width. The `<div>` elements have little content—just the numbers 1, 2 and 3—yet their entire area is visible because of the background color.

```
<!DOCTYPE html>
<html>
    <head>
        <title>CSS box size</title>
        <style>
            div {
                height: 100px;
                width: 20%;
                background-color: powderblue;
            }
        </style>
    </head>
```

```
    <body>
        <p>
            The following three 'div' elements have heights of
            <b>100px</b> and widths of <b>20%:</b>
        </p>
        <div>1</div>
        <div>2</div>
        <div>3</div>
    </body>
</html>
```

The resulting page is shown in Figure 2.6.

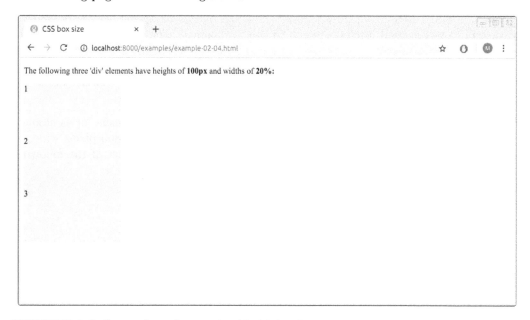

FIGURE 2.6: Screenshot of `example-02-04.html`

2.8.4.3   `border-width:`, `border-style:`, `border-color:`, `border-radius:`

These properties control the appearance of **borders** (Figure 2.5) around an element. The `border-width:` property sets border width. The `border-style:` property sets border style. Possible values for `border-style:` include `solid`, `dotted`, `double` and `dashed`. The `border-color:` property sets border color.

The `border-width:`, `border-style:` and `border-color:` properties affect all four borders. There are also specific properties that affect only the top, left, right, or bottom border of an element. These can be used by adding `-top-`, `-left-`, `-right-`, or `-bottom-` in the property name, respectively. For example, the following CSS rule can be used to produce horizontal lines at the top of all paragraphs on a web page:

```
p {
    border-top-width: 1px;
    border-top-style: solid;
}
```

The `border-radius:` property can be used to make the border corners appear **rounded** (see Section 2.10 for an example).

2.8.4.4   `margin:`

The `margin:` property controls the **margin**—how much space is allocated around the *outside* of the element (Figure 2.5), that is, between the given element and its neighboring elements. The size of margins can be expressed using different units of length, same as in the `width:` and `height:` properties (Section 2.8.4.2).

The `margin:` property affects all four margins. Just like with borders (Section 2.8.4.3), there are also properties for controlling individual margins: `margin-top:`, `margin-bottom:`, `margin-right:`, and `margin-left:`.

2.8.4.5   `padding:`

The `padding:` property controls box **padding**—how much space is allocated between the border of the element and its contents (Figure 2.5). Values are specified the same way as for margins. Similarly, there are specific properties (`padding-top:`, etc.), for controlling the padding on each side of the element.

- Make your own copy of `example-02-04.html` and modify the following aspects:
    - Add more text inside the `<div>` elements.
    - Add CSS rules to show a colored border around each `<div>` element.
    - Add CSS rules to separate the `<div>` elements from each other (using margins) and to separate their text from the border (using padding).

2.8.4.6   `position:`, `top:`, `bottom:`, `right:`, `left:`

The `position:` property specifies the **positioning method** for an element. Possible values for the `position:` property include:

- `static` (the default)
- `relative`
- `absolute`
- `fixed`

In **normal** flow (`position: static`), each block-level element sits below the previous one. Since this is the default way in which browsers treat HTML elements, you do not need a CSS property to indicate that elements should appear in normal flow, but the syntax would be:

```
position: static;
```

Relative positioning with `position: relative` moves an element in relation to where it would have been in normal flow. For example, you can move it `10px` lower than it would have been in normal flow or 20% to the right. With relative positioning, the offset properties, `top:` or `bottom:` and/or `left:` or `right:`, are used to indicate how far to move the element from where it would have been in normal flow. To move the box up or down, you can use the `bottom:` or `top:` offset properties, respectively. To move the box left of right, you can use the `right:` or `left:` offset properties, respectively. The values of the box offset properties are usually given in pixels (`10px`), percentages (`30%`), or ems (`8em`).

Absolute positioning with `position: absolute` places the element relatively to the nearest "positioned" ancestor. A "positioned" element is one whose position is anything except `position: static`. However, if an absolute positioned element has no positioned ancestors, it uses the document `<body>`. For example, we can specify that a `<div>` is located 10 pixels from the top and 10 pixels from the left of its containing `<div>`. An absolutely positioned element moves along with page scrolling, i.e., does not stay in the same place on screen.

Fixed positioning with `position: fixed` is relative to the viewport. As implied by the name *fixed*, this means the element always stays in the same place even if the page is scrolled.

It is important to note that both `position: absolute` and the `position: fixed` positioned elements are *taken out* of normal flow. This means the elements "hover" above the page, while other elements shift to fill the place they previously occupied. This also means that these elements can obstruct the view of other elements.

The `example-02-05.html` web page (Figure 2.7) has three `<div>` elements, styled to have specific dimensions, background color, and borders, using the CSS code shown below:

```
div {
    height: 100px;
    width: 150px;
    background-color: powderblue;
    padding: 20px;
    margin: 5px;
    border-width: 1px;
    border-style: solid;
}
```

There are two additional styling rules, `div#one` and `div#two` (see below), affecting just the first and second `<div>` elements, respectively. The first `<div>` has a relative position, thus shifted from its normal flow by **30px** from bottom and by **40px** from left. Note that though the element is shifted, its original position is maintained in normal flow, rather than filled up with other content (Figure 2.7).

```
div#one {
    position: relative;
    bottom: 30px;
    left: 40px;
}
```

The second `<div>` element has an absolute position, placing it relative to the nearest "positioned" ancestor. Since it has no such ancestors, the `<div>` is placed relative to the `<body>`, that is, relative to the viewport. Specifically, the second `<div>` is placed **30px** from the bottom of the viewport and **40px** from the left of the viewport. Note that the second `<div>` obstructs some of the text shown in the `<p>` elements, as well as the third `<div>`.

Both of these outcomes are because absolutely positioned elements go out of the ordinary flow. Therefore the second `<div>` is not affected by underlying text, as if it is "hovering" above the text. The place in the normal flow is then cleared for other content to take place, such as the third `<div>` (Figure 2.7).

```
div#two {
    position: absolute;
    bottom: 30px;
    left: 40px;
}
```

The resulting page is shown in Figure 2.7.

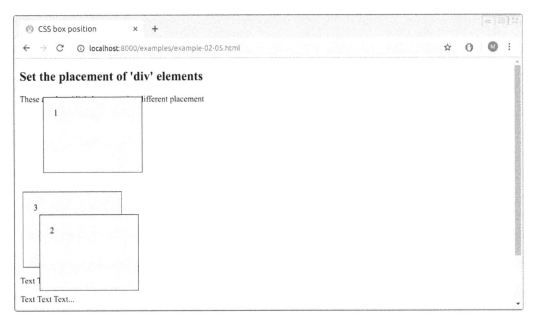

FIGURE 2.7: Screenshot of `example-02-05.html`

- Create your own copy of `example-02-05.html`.
- Delete the `div#one` rule from the source code.
- Open the new version in the browser to observe how the first `<div>` goes back to its "natural" position (compared to Figure 2.7).

2.8.4.7   z-index:

As demonstrated in `example-02-05.html`, moving elements out of normal flow can lead to *overlap* between boxes (Figure 2.7). The `z-index:` property allows you to control which box appears on top.

The `z-index` value can be specified with any integer (positive, zero, or negative). An element with greater `z-index` value is always in front of an element with a lower `z-value`. Note that the `z-index` only works on *positioned* elements, i.e., those positioned with `position: relative`, `position: absolute`, or `position: fixed`[13].

2.9   Hurricane scale example

In this section and the next one, we will summarize what we learned about CSS in two practical examples. In the first example, we create a page that visualizes the hurricane scale based on wind speed[14], going from a tropical depression ($\leq 17 \frac{m}{s}$) to a level-five hurricane ($\geq 70 \frac{m}{s}$). We begin with a plain HTML code of `example-02-06.html`, without any styling:

```
<!DOCTYPE html>
<html>
    <head>
        <title>Hurricane scale</title>
    </head>
    <body>
        <h1>Hurricane Scale</h1>
        <p id="five">Five &ge;70 m/s</p>
        <p id="four">Four 58-70 m/s</p>
        <p id="three">Three 50-58 m/s</p>
        <p id="two">Two 43-49 m/s</p>
        <p id="one">One 33-42 m/s</p>
        <p id="ts">Tropical storm 18-32 m/s</p>
        <p id="td">Tropical depression &le;17 m/s</p>
        <p>
            <a href="https://en.wikipedia.org/wiki/
                Saffir%E2%80%93Simpson_scale">(Data Source)
            </a>
        </p>
    </body>
</html>
```

This page has one `<h1>` heading and seven `<p>` paragraphs representing the various wind-speed classes. Note that `&le;` and `&ge;` are special ways to specify the $\leq$ and $\geq$ characters[15], respectively, using HTML code. The result does not look very impressive (Figure 2.8).

---

[13]More information and examples on positioning can be found in the *Position* article (https://developer.mozilla.org/en-US/docs/Web/CSS/position) by Mozilla.

[14]https://en.wikipedia.org/wiki/Saffir%E2%80%93Simpson_scale

[15]https://www.w3schools.com/charsets/ref_utf_math.asp

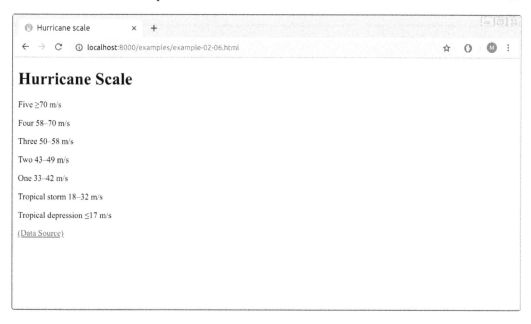

FIGURE 2.8: Screenshot of `example-02-06.html`

We can improve this web page by adding graduated colors to the hurricane classes. Here is CSS code that we can use to make the page look more interesting:

```
body {
  background-color: silver;
  padding: 20px;
  font-family: Arial, Verdana, sans-serif;
}
h1 {
  background-color: rgba(255,255,255,0.5);
  color: #64645A;
  padding: inherit;
}
p {
  padding: 5px;
  margin: 0px;
}
p#five { background-color: #ff6060; }
p#four { background-color: #ff8f20; }
p#three { background-color: #ffc140; }
p#two { background-color: #ffe775; }
p#one { background-color: #ffffcc; }
p#ts { background-color: #00faf4; }
p#td { background-color: #5ebaff; }
```

- Edit a copy of **example-02-06.html** and link the previous CSS code to the HTML document. You can use either of the last two methods presented in Section 2.7:
  - Embedded CSS—Code goes in the **<style>** element within the **<head>** element of the HTML document (Section 2.7.3)
  - External CSS—Code goes in a separate **.css** file, then linked using the **<link>** element within the **<head>** element of the HTML document (Section 2.7.4)
- Remove or modify some of the rules to examine their effect.

The hurricane-scale page, after applying the CSS rules, is given in **example-02-07.html** (Figure 2.9).

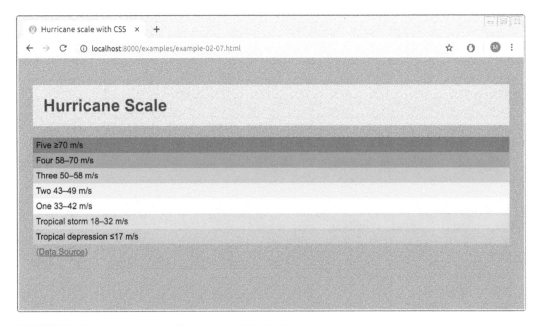

FIGURE 2.9: Screenshot of **example-02-07.html**

2.10   Map description example

In our second practical CSS example, we are going to create styled title and description panels, which can be used to describe a web map in cases when the map fills the entire screen. You can see the final result we aim at in Figure 2.13.

We start with an HTML document that has the following content in its `<body>` element:

```html
<h1>Leaflet Quick Start Guide</h1>
<div id="description">
    <h2>About this map</h2>
    <p>This is the final result of the
        <a href="https://leafletjs.com/examples/quick-start/"
            target="_blank">Leaflet Quick Start Guide</a>.
    The map demonstrates Leaflet basics, such as setting up a Leaflet map,
    adding markers, polylines and popups.</p>
    <p>The placement and styling of the title and description boxes is
    done using CSS. Check out the source code of this page to see how.</p>
    <p>Map authored by Michael Dorman</p>
</div>
```

The HTML code contains:

- An `<h1>` element, which will comprise the map *title*
- A `<div>` element, which will comprise the map *description*

The map *title* in the `<h1>` element is just text, saying `Leaflet Quick Start Guide`. The map *description* in the `<div>` element is composed of a smaller title (`<h2>`) and three paragraphs (`<p>`). Using the `<div>` element we bind the entire description into a group (Section 1.6.11.2) with a specific ID, `id="description"`. The ID will be used to set the description style with CSS. It is not necessary to set an ID for the map title `<h1>` element, since the page has only one `<h1>` element anyway, therefore it can be targeted with a type selector (Section 2.4.2).

The way our HTML page appears—before any CSS styling is applied—is shown in Figure 2.10. This is quite different from the final result (Figure 2.13), but hang on. We will get there using CSS styling.

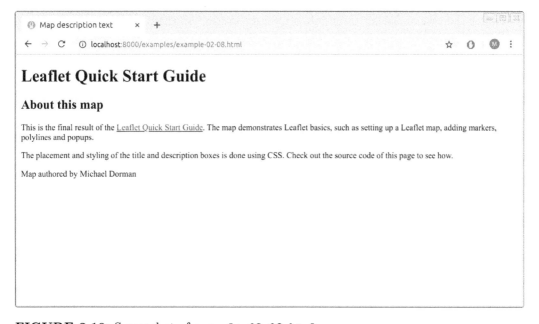

FIGURE 2.10: Screenshot of `example-02-08.html`

Moving on to styling. Assuming our web map is going to fill the entire screen, we would like to position the title and description in boxes "above" the map. As discussed in Section 2.8.4.6, this means we need the boxes to exit normal flow, using `position: absolute` or `position: fixed`. To achieve the modified positioning, as well as specify other aesthetic properties of the boxes (discussed below), we add the following CSS rules in `example-02-09.html`:

```css
h1 {
    position: absolute;
    top: 10px;
    left: 55px;
    margin-top: 0;
    padding: 10px 15px;
    background-color: rgba(255, 255, 255, .5);
    border: 1px solid grey;
    border-radius: 3px;
}
#description {
    position: absolute;
    bottom: 0;
    left: 10px;
    width: 280px;
    margin-bottom: 20px;
    padding: 0 15px;
    background-color: rgba(255, 255, 255, .7);
    border: 1px solid grey;
    border-radius: 3px;
}
```

In the above CSS code, we are using several concepts:

- The `position: absolute` property along with `top:` and `left:` or `bottom:` and `left:`, to put the title and description into a specific position on the page
- The `margin-top:`, `margin-bottom`, and `padding:` properties, to make sure there is enough space around the text as well as around the boxes
- The `border:` and `border-radius:`, to make the box border visible and having rounded corners
- The `background-color:` property to make the background color semi-transparent white

There are also two CSS *shortcuts*, which we have not met up until now. CSS has many types of shortcuts of this sort, and we will encounter several other ones in later chapters:

- Using `border: 1px solid grey;` as a shortcut for `border-width: 1px`, `border-style: solid` and `border-color: grey` all in one declaration
- Using `padding: 10px 15px;` as a shortcut for `padding-top: 10px`, `padding-bottom: 10px`, `padding-right: 15px` and `padding-left: 15px` in one declaration

The result after applying the above CSS rules is shown in Figure 2.11. As you can see, the title and map description are now in separate bordered boxes, positioned in the specified locations inside the viewport.

There are more CSS properties that can be modified to make the boxes even nicer. For example, we can use *specific fonts* (free ones, loaded from a Google server) and change the *link style* when hovering with the mouse. We are not going to learn about these properties,

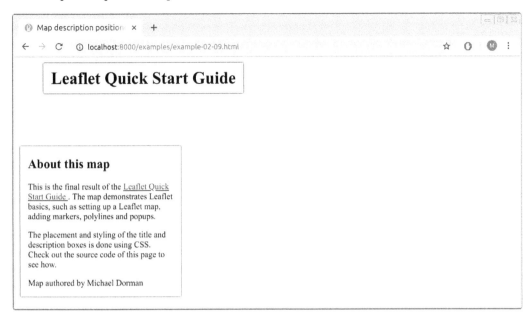

FIGURE 2.11: Screenshot of `example-02-09.html`

but you are welcome to explore them in `example-02-10.html`, which gives the slightly modified result shown in Figure 2.12.

FIGURE 2.12: Screenshot of `example-02-10.html`

Finally, we are missing the web map itself which the title and description refer to. The final version `example-02-11.html` adds a web map in the background. If you inspect its source code, you will see that one of the things that has been added in the code is the following `<script>` element (Section 1.6.3.3) in the end of the `<body>`:

```
<script src="map.js"></script>
```

This `<script>` element loads a JavaScript code from a separate file named `map.js`. Another thing that was added in the HTML code is a `<div>` element that has `id="map"`. The script acts on the `<div>` element that has `id="map"`, replacing it with an interactive map. The content of the script file `map.js` is not important for now, as we will learn all about it in Chapter 6.

Note that another addition in `example-02-11.html` is that both the map title and the map description are associated with the following `z-index` declaration, to make sure they are shown *above* the map (Section 2.8.4.7):

```
z-index: 800;
```

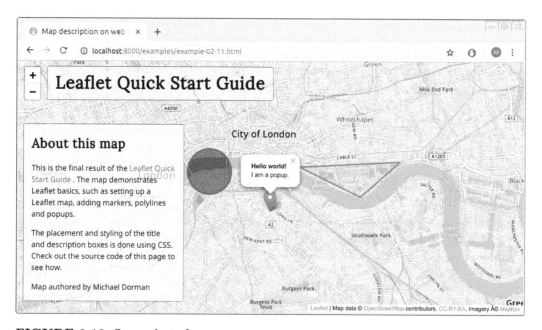

FIGURE 2.13: Screenshot of `example-02-11.html`

2.11   Exercise

- Think about the data and concept that you would like to use for a web-mapping project of your own.
- Modify the content of the map description from the last example to describe your planned map.

3

JavaScript Basics

3.1 Introduction

So far we learned how to define the contents and style of web pages, with HTML (Chapter 1) and CSS (Chapter 2), respectively. We now move to the third major web technology, **JavaScript**. Being a programming language, JavaScript is fundamentally different from HTML and CSS, which are essentially data formats. Using HTML and CSS we can instruct the browser to display the given contents, with particular styling. Using a programming language, such as JavaScript, we can give the browser (or other computing environments) a much wider variety of instructions. For example, we can instruct the computer to perform mathematical calculations, to show or hide contents in response to user input, to periodically load up-to-date contents from an external source, and so on.

JavaScript is the most important of the technologies presented in this book, and we will use it extensively throughout the later chapters to build interactive web maps. Chapter 3 (this chapter) introduces the basic principles of the JavaScript language and the practical way of experimenting with it, using the JavaScript console built into web browsers (Figure 3.1). All of the examples we will see in this chapter work "in isolation," not affecting web page content and style. Therefore, while reading this chapter, the reader can temporarily put aside what we learned in Chapters 1–2. Then, in Chapter 4, we will go back to HTML and CSS and see how to associate our JavaScript code with the web page contents, and how to modify that contents in response to user input. Finally, in Chapters 6–13, we will apply these techniques in the context of web mapping, using JavaScript to create customized web maps and to control their behavior.

3.2 What is JavaScript?

JavaScript[1] is a programming language, mostly used to define the interactive behavior of web pages. JavaScript allows you to make web pages more interactive by accessing and modifying the contents and styling in a web page while it is being viewed in the browser. JavaScript, along with HTML and CSS, forms the foundation of modern web browsers and the internet.

JavaScript is considered, unofficially, to be the most popular programming language in the world (Figure 2). Numerous open-source mapping and data visualization libraries are written in JavaScript, including **Leaflet**, which we use in this book, **OpenLayers**, **D3**, and many

---

[1]https://en.wikipedia.org/wiki/JavaScript

others. Many commercial services of web mapping also provide a JavaScript API for building web maps with their tools, such as **Google Maps JavaScript API**, **Mapbox-GL-JS** and **ArcGIS API for JavaScript** (more on that in Section 6.4).

3.3   Client-side vs. server-side

When talking about programming for the web, one of the major concepts is the distinction between **server-side** and **client-side**.

- The **server** is the location on the web that serves your website to the rest of the world.
- The **client** is the computer that is accessing that website, requesting information from the server.

In server-side programming, we are writing scripts that run on the *server*. In client-side programming, we are writing scripts that run on the *client*, that is—in the browser. This book focuses on *client-side* programming, though we will say a few words on server-side programming in Chapter 5. JavaScript can be used in both server-side[2] and client-side programming, but is primarily a client-side language, working in the browser on the client computer.

As we will see later on, there are two fundamental ways for using JavaScript on the client-side:

- Executing scripts when the web page is being **loaded**, such as scripts that instruct the browser to download data from an external location and display them on the page
- Executing scripts in response to user **input** or interaction with the page, such as performing a calculation and showing the result when the user clicks on a button

When JavaScript code is executed in a web page, it can do many different types of things, such as modifying the contents of the page, changing the appearance of content, sending information to a server, or getting information from the server.

3.4   The JavaScript console

When loading a web page which is associated with one or more scripts, the JavaScript code is automatically *interpreted* and executed by the browser. We can also manually interact with the browser's interpreter or engine[3] using the JavaScript **console**, also known as the command line, which is built into all modern web browsers. The console is basically a way to type code into a computer and get the computer's answer back. This is useful for experimenting with JavaScript code, such as quickly testing short JavaScript code snippets. It is exactly what we are going to do in this chapter.

---

[2]**Node.js** (https://nodejs.org/en/) is a prominent example of a *server-side* environment for executing JavaScript code.

[3]https://en.wikipedia.org/wiki/JavaScript_engine

- Open a web browser.
- In Chrome: open the JavaScript console by pressing **Ctrl+Shift+J**, or by opening the developer tools with **F12**, then clicking on the **Console** tab.
- Type **5+5** and press **Enter**.
- You have just typed a JavaScript expression, which was interpreted and the returned value 10 was printed.
- The expression sent to the interpreter is marked with the > symbol. The returned value being printed is marked with the <- symbol (Figure 3.1).
- Use **Ctrl+L** to clear the console.

Figure 3.1 shows a screenshot of the JavaScript console with the entered expression **5+5** and the response **10**.

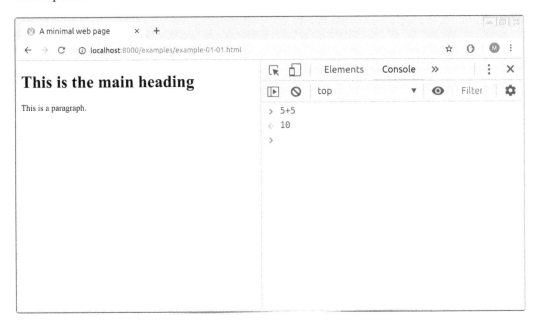

FIGURE 3.1: The JavaScript console in Chrome

In this chapter, we will be using the console to get familiar with the JavaScript language. We will be typing expressions and examining the computer's responses. In the following Chapters 4–13, instead of typing our JavaScript code, we will mostly load the code using the `<script>` element (Section 1.6.3), either from embedded scripts or from external files. Nevertheless, the console is still going to be useful to interactively explore function behavior or currently loaded objects, or to interactively **debug**[4] our scripts when they contain errors.

---

[4]https://en.wikipedia.org/wiki/Debugging

3.5   Assignment

Assignment to variables is one of the most important concepts in programming, since it lets us keep previously calculated results in memory for further processing. **Variables** are containers that hold data values, simple or complex, that can be referred to later in your code.

In JavaScript, variables are **defined** using the keyword `var`, followed by the variable name. For example, the following two expressions define the variables x and y:

```
var x;
var y;
```

Note that each JavaScript expression should end with `;`. Ending statements with semicolon is not strictly required, since statement ending can also be automatically determined by the browser, but it is highly recommended to use `;` in order to reduce ambiguity.

Values can be **assigned** to variables using the assignment operator `=`. Assignment can be made along with variable definition:

```
var x = 5;
var y = 6;
```

or later on, in separate expressions:

```
x = 5;
y = 6;
```

Either way, we have now defined two variables x and y. The values contained in these two variables are **5** and **6**, respectively. In your JavaScript console, to see a current value of a variable, type its name and hit **Enter**. The current value will be returned and printed:

```
x;   // Returns 5
```

The above code uses a JavaScript **comment**. We already learned how to write HTML comments (Section 1.5.2) and CSS comments (Section 2.8.2.1) in previous chapters. In JavaScript:

- Single line comments start with `//`
- Single or multi-line comments start with `/*` and end with `*/` (like in CSS)

There are several short-hand assignment operators, in addition to the basic assignment operator `=`. For example, another commonly used assignment operator is the **addition assignment** operator `+=`, which adds a value to a variable. The following expression uses the addition assignment operator `+=`:

```
x += 10;
```

This is basically a shortcut for the following assignment[5]:

```
x = x + 10;
```

3.6 Data types

3.6.1 Overview

In the examples in Section 3.5 we defined variables holding *numeric* data, such as the numbers 5 and 6. JavaScript variables can also hold other **data types**. JavaScript data types are divided into two groups: **primitive** data types and **objects**.

Primitive data types include the following:

- **String**—Text characters that are delimited by quotes, for example: `"Hello"` or `'Hello'`
- **Number**—Integer or floating-point value, for example: `5` and `-10.2`
- **Boolean**—Logical values, `true` or `false`
- **Undefined**—Specifies that a variable is declared but has no defined value, `undefined`
- **Null**—Specifies an intentional absence of any object value, `null`

Objects are basically collections of the primitive data types and/or other objects:

- **Array**—Multiple variables in a single variable, for example: `["Saab", "Volvo", "BMW"]`
- **Object**—Collections of `name:value` pairs, for example: `{type:"Fiat", model:"500", color:"white"}`

The data types in JavaScript are summarized in Table 3.1. In the following Sections 3.6.2–3.6.4 we go over the data types in JavaScript one by one.

TABLE 3.1: JavaScript data types

Group	Data type	Example
Primitive	String	`"Hello"`
	Number	`5`
	Boolean	`true`
	Undefined	`undefined`
	Null	`null`
Objects	Array	`["Saab", "Volvo"]`
	Object	`{type: "Fiat", model: "500"}`

---

[5] More information on `+=`, and other assignment operators, can be found in the *Assignment Operators* reference (`https://developer.mozilla.org/en-US/docs/Web/JavaScript/Reference/Operators/Assignment_Operators`) by Mozilla.

3.6.2   Primitive data types

3.6.2.1   String

Strings are collections of text characters, inside single (') or double (") quotes[6]. For example, the following expressions define two variables that contain strings, s1 and s2:

```
var s1 = 'Hello';
var s2 = "World";
```

Strings can be **concatenated** using the + operator:

```
s1 + s2;         // Returns "HelloWorld"
s1 + " " + s2;   // Returns "Hello World"
```

3.6.2.2   Number

Numbers can be integers or decimals. The usual **arithmetic operators**, +, -, *, and /, can be used with numeric variables:

```
var x = 5, y = 6, z;
z = x + y;  // Returns 11
```

In this example, we defined two variables x and y and assigned their values (5 and 6). We also defined a third variable z without assigning a value. Note how we defined x, y, and z in the same expression, separated by commas—this saves typing the var keyword three times. In the last expression we calculated x+y and assigned the result of that calculation into z.

JavaScript also has **increment** ++ and **decrement** -- operators. The increment operator ++ adds one to the current number. The decrement operator -- subtracts one from the current number. For example:

```
var x = 5;
x++;  // Same as: x=x+1; the value of x is now 6
x--;  // Same as: x=x-1; the value of x is now 5
```

When running the last two expressions, you will see 5 and 6 (not 6 and 5) printed in the console, respectively. This is because the increment ++ and decrement -- expressions return the current value, before modifying it[7].

---

[6]For more information on strings, see the *Strings* reference (https://developer.mozilla.org/en-US/docs/Learn/JavaScript/First_steps/Strings) by Mozilla.

[7]For more information on numbers, see the *Numbers and Operators* reference (https://developer.mozilla.org/en-US/docs/Learn/JavaScript/First_steps/Math) by Mozilla.

3.6.2.3 Boolean

Boolean (logical) values are either **true** or **false**. For example:

```
var found = true;
var lost = false;
```

In practice, Boolean values are usually created as the result of **comparison operators**. For example:

```
9 >= 10;   // Returns false
11 > 10;   // Returns true
```

JavaScript comparison operators are summarized in Table 3.2.

TABLE 3.2: JavaScript comparison operators

Operator	Meaning
==	Equal to
===	Equal value and equal type
!=	Not equal
!==	Not equal value or not equal type
>	Greater than
<	Less than
>=	Greater than or equal to
<=	Less than or equal to

You may have noticed there are two versions of the equality (==, ===) and inequality (!=, !==) comparison operators (Table 3.2). The difference between them is that the former ones (==, !=) consider just the value, while the latter ones (===, !==) consider both value and type. What do the terms *equal value* and *equal type* mean? Consider the following example, where 5 and "5" have equal value (5) but unequal type (number vs. string). Comparing these two values gives different results when using the == and === operators. This is because == considers just the value, while === is more strict and considers both value and type:

```
"5" == 5;    // Returns true
"5" === 5;   // Returns false
```

Since the conversion rules of the == and != operators are complicated and unmemorable, it is not advised to use them. Instead, the === and !== operators, with their expected and strict behaviors, are recommended (Crockford, 2008).

Boolean values can be combined with **logical operators**:

- &&—**AND** operator, **true** only if both values are **true**
- ||—**OR** operator, **true** if at least one of the values is **true**[8]
- !—**NOT** operator, **true** if value is **false**

---

[8]Note the distinction between && and &, and between || and |. The latter, & and |, are known as *bitwise operators* and are not intended for combining comparison expressions.

For example:

```
var x = 6;
var y = 3;
(x < 10 && y > 1);    // Returns true
(x == 5 || y == 5);   // Returns false
!(x == 5 || y == 5);  // Returns true
```

Boolean values are commonly used for flow control, which we learn about later on (Section 3.10).

3.6.2.4   Undefined

Undefined (`undefined`) means that a variable has been declared but has not been assigned with a value yet. For example:

```
var x;
x;  // Returns undefined
```

3.6.2.5   Null

Null (`null`) is another special data type, representing lack of a value. For example:

```
var x = null;
```

3.6.3   Objects

3.6.3.1   Array

JavaScript **arrays** are used to store an ordered set of values in a single variable. An array can be created by putting zero or more values inside a pair of square brackets ([and]), separating the values with commas (,). For example:

```
var a = [];          // Empty array
var b = [1, 3, 5];   // Array with three elements
```

Array items can be accessed using the brackets ([) along with a numeric **index**. Note that index position in JavaScript starts at 0. For example:

```
b[0];  // Returns 1
b[1];  // Returns 3
b[2];  // Returns 5
```

An array can be composed of different types of values, though this is rarely useful:

```
var c = ["hello", 10, undefined];
```

An array element can also be an array, or an object (Section 3.6.3.2 below). For example, an array of arrays can be thought of as a **multi-dimensional** array:

```
var d = [[1, 2, 3], [4, 5, 6], [7, 8, 9]];
```

To access an element inside a multi-dimensional array we can combine several array access operations:

```
d[2][1];  // Returns 8
```

In the last expression, we used `d[2]` to access the element in position 2 of `d`, which is `[7, 8, 9]`. This is actually the *third* element of `d`, since, as you remember, JavaScript index positions start at 0. Then, we used `d[2][1]` to access the *second* element of `[7, 8, 9]`, which gives us the final value of 8[9].

3.6.3.2   Objects

JavaScript **objects** are collections of named values[10]. An object can be defined using curly brackets (`{` and `}`). Inside the brackets, there is a list of `name: value` pairs, separated by commas (`,`). For example:

```
var person = {
    firstname: "John",
    lastname: "Smith",
    age: 50,
    eyecolor: "blue"
};
```

The above expression defines an object named `person`. This object is composed of four named values, separated by commas. Each named value is composed of a name (such as `firstname`) and a value (such as `"John"`), separated by a colon (`:`).

The named values are also called object **properties**. For example, the above object has four properties, named `firstname`, `lastname`, `age`, and `eyecolor`. The property values can be of any data type, including primitive data types (as in the above example, `"John"`, `"Smith"`, 50, `"blue"`), but also arrays and objects. Property values can also be functions, in which case they are referred to as *methods* (Section 3.8).

Objects are fundamental to JavaScript, and almost everything we work with in JavaScript is an object. The rationale of an object is to bind related data and/or functionality into a single collection. The collection usually consists of several variables and functions, which are called properties and methods when they are inside objects, respectively.

A JavaScript object is comparable to a real-life object. For example, a car can be thought of as an object (Figure 3.2). The car has properties like weight and color, which are set to certain values, and it has methods, like starting and stopping.

---

[9]For more information on arrays, see the *Arrays* article (`https://developer.mozilla.org/en-US/docs/Learn/JavaScript/First_steps/Arrays`) by Mozilla.

[10]JavaScript objects are comparable, for example, to *dictionaries* in Python or to *lists* in R.

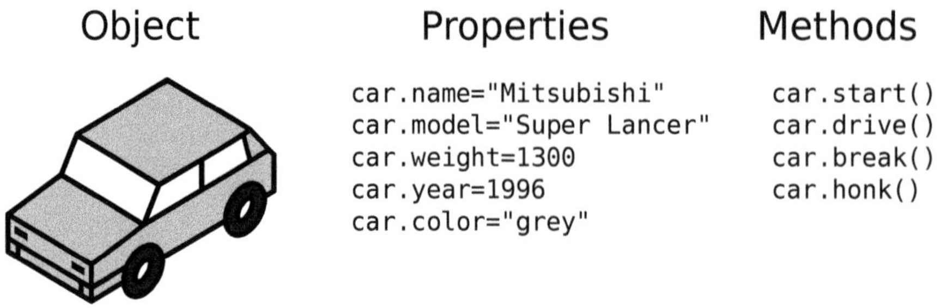

Object   Properties   Methods

```
car.name="Mitsubishi"         car.start()
car.model="Super Lancer"      car.drive()
car.weight=1300               car.break()
car.year=1996                 car.honk()
car.color="grey"
```

FIGURE 3.2: A JavaScript object has properties and methods, just like a real-life object

Object properties can be accessed using either of the following two methods:

- The **dot** notation (`.`)
- The **bracket** notation (`[`)

In both cases, we need to specify the object name and the property/method name. For example, getting the **person** properties using the dot notation looks like this:

```
person.firstname;   // Returns "John"
person.age;         // Returns 50
person.firstname + " is " + person.age + " years old.";
```

- What do you think will be returned by the third expression in the above code section?
- Create the **persons** object in the console and run the expression to check your answer.

The same can be accomplished using the bracket notation, as follows:

```
person["firstname"];   // Returns "John"
person["age"];         // Returns 50
person["firstname"] + " is " + person["age"] + " years old.";
```

When using the bracket notation, property names are specified as strings, in quotes. This means the dot notation is shorter to type, but the bracket notation is useful when the property name is specified with a variable:

```
var property = "firstname";
person[property];   // This works
person.property;    // This doesn't work
```

As we already mentioned, an object property value can be another object or an array rather than a primitive data type. For example, we could arrange the information in the **person** object in a different way:

```
var person = {
    name: {firstname: "John", lastname: "Smith"},
    age: 50,
    eyecolor: "blue"
};
```

Instead of having individual **firstname** and **lastname** properties, we now have a **name** property which is an *internal* object, containing the **firstname** and **lastname** properties. To reach an internal object within another object, the dot notation (or the bracket notation) can be repeated several times in the same expression:

```
person.name.firstname;   // Returns "John"
```

This is typical syntax that you will see a lot in JavaScript code:

```
object1.object2.property;
```

- Note the **auto-complete** functionality, which can make it easier to interactively construct this kind of expression interactively in the console.
- For example, create the **person** object in the console, then start typing **person.** or **person.name.** to observe the auto-complete suggestions.

Object properties can also be modified via assignment. For example, we can change the person name from **"John"** to **"Joe"** as follows[11]:

```
person.name.firstname = "Joe";
person.name.firstname;   // Returns "Joe"
```

3.6.4 Checking type of variables

The **typeof** operator can always be used to query the type of variable we are working with. The following expressions demonstrate the use of **typeof** on primitive data types:

---

[11]For more information on objects, see the *Objects* reference (`https://developer.mozilla.org/en-US/docs/Learn/JavaScript/Objects/Basics`) by Mozilla.

```
typeof "a";         // Returns "string"
typeof 1;           // Returns "number"
typeof false;       // Returns "boolean"
typeof undefined;   // Returns "undefined"
typeof null;        // Returns "object" (!)
```

Note that `typeof` returns `"object"` when applied to `null`, even though `null` is a primitive data type. This behavior is considered to be a bug[12] in the JavaScript language, kept for legacy reasons.

Arrays and objects are collectively classified as *objects*:

```
typeof [1,5];    // Returns "object"
typeof {a: 1};   // Returns "object"
```

3.7   Functions

A **function** is a block of code designed to perform a particular task. If different parts of our JavaScript code need to perform the same task, we do not need to repeat the same code block multiple times. Instead, we *define* a function once, then *call* it multiple times whenever necessary.

A function is *defined* with:

- The `function` keyword
- A function name of our choice, such as `multiply`
- Parameter names separated by commas, inside parentheses (`(` and `)`), such as `(a, b)`
- The code to be executed by the function, curly brackets (`{` and `}`), such as `{return a * b;}`

The function code may contain one or more expressions. One or more of those can contain the `return` keyword, followed by a value the function **returns** when executed. When a `return` expression is reached the function stops, and the value after `return` is returned by the function. In case there is no `return` expression in the function code, the function returns `undefined`.

For example, the following expression **defines** a function named `multiply`. The `multiply` function has two **parameters**, `a` and `b`. The function returns the result of `a` multiplied by `b`.

```
// Function definition
function multiply(a, b) {
    return a * b;
}
```

---

[12]http://2ality.com/2013/10/typeof-null.html

Once the function is defined, you can execute it with specific **arguments**. This is known as a function **call**. For example, the following expression is a function call of the `multiply` function, returning 20:

```
// Function call
multiply(4, 5);
```

Note the distinction between parameters and arguments. Function *parameters* are the names listed in the function definition. Function *arguments* are the real values passed to the function when it is called. In the above example, the parameters of the `multiply` function are `a` and `b`. The arguments in the above function call of `multiply` were 4 and 5.

A function does not necessarily have any parameters at all, and it does not necessarily return any value. For example, here is a function that has no parameters and does not return any value[13]:

```
function greeting() {
  console.log("Hello!");
}
```

The code in the `greeting` function uses the `console.log` function, which we have not met until now. The `console.log` function prints text into the console.

```
greeting();  // Prints "Hello!" in the console
```

The `console.log` function is very helpful when experimenting with JavaScript code, and we will use it often. For example, when running a long script we may wish the current value of a variable to be printed out, to monitor its change through our program.

To keep the returned value of a function for future use, we use assignment[14]:

```
var x;
x = multiply(4, 5);
```

3.8   Methods

3.8.1   What are methods?

When discussing objects (Section 3.6.3.2 above), we mentioned that an object property can also be a function, in which case it is also known as a **method**. For example, a method may be defined when creating an object, in case one of the properties is assigned with a function definition:

---

[13] As mentioned above, a function with no `return` statement returns the default value of `undefined`.

[14] For more information on functions, see the *Functions* article (`https://developer.mozilla.org/en-US/docs/Learn/JavaScript/Building_blocks/Functions`) by Mozilla.

```
var objectName = {
  ...,
  methodName: function() { expressions; },   // Method definition
  ...
}
```

Note that, in this example, the function was defined a little differently—there was no function name between the **function** keyword and the list of parameters (). This type of function is called an **anonymous function**. Once it is defined, a method can be accessed just like any other property:

```
objectName.methodName();   // Method access
```

The parentheses () at the end of the last expressions imply we are making a function call for the method. For example, let us define a **car** object (Figure 3.2), which has several properties with primitive data types, and one method named **start**:

```
var car = {
    name: "Mitsubishi",
    model: "Super Lancer",
    weight: 1300,
    year: 1996,
    color: "grey",
    start: function() { console.log("Car started!"); }
};
```

Generally, methods define the actions that can be performed on objects. Using the car as an analogy, the methods might be *start*, *drive*, *brake*, and *stop*. These are actions that the car can perform. For example, *starting* the car can be done like this:

```
car.start();   // Prints "Car started!"
```

As a more realistic example, think of an object representing a layer in an interactive web map (Section 6.6). The layer object may have properties, such as the layer geometry, symbology, etc. The layer object may also have methods, such as methods for adding it to or removing it from a given map, adding popup labels on top of it, returning its data in a different format, and so on.

- Add another method to the **car** object, named **stop**, which prints a "Car stopped!" message in the console.
- Try using the new **stop** method with **car.stop()**.

3.8.2 Array methods

In JavaScript, primitive data types and arrays also have several predefined properties and methods, even though they are not strictly objects[15]. In this section, we look into a few useful properties and methods of arrays: `.length`, `.pop`, and `.push`.

The `.length` property of an array gives the number of items that it has:

```
var a = [1, 7, [3, 4]];
a.length;      // Returns 3
a[2].length;   // Returns 2
```

- In the above code section:
  - Why does the first expression return 3?
  - Why does the second expression return 2?

The `.pop` and `.push` methods can be used to remove or to add an item at the end of an array, respectively. The `.pop` method removes the last element from an array. For example:

```
var fruits = ["Orange", "Banana", "Mango", "Apple"];
fruits.pop();  // Removes the last element from fruits
```

Note that the `.pop` method, like many other JavaScript methods, modifies the array *itself*, rather than creating and returning a modified copy of it. The returned value by the `.pop` method is in fact the item which was removed. In this case, the returned value is `"Apple"`. In other words, when executing the above expressions the value `"Apple"` is printed, while the new value of `fruits` becomes `["Orange", "Banana", "Mango"]`.

The `.push` method adds a new element at the end of an array:

```
var fruits = ["Orange", "Banana", "Mango", "Apple"];
fruits.push("Pineapple");  // Adds a new element to fruits
```

Again, note that the `.push` method modifies the original array, which means that `fruits` now has length of 5 (including `"Pineapple"` at the end), after the above expressions are executed. The returned value of the `.push` method is the new array length. In this example, the returned value is 5.

---

[15]The mechanism which makes this happen is beyond the scope of this book.

3.9   Scope

Variable **scope**[16] is the region of a computer program where that variable is accessible or "visible". In JavaScript there are two types of scope:

- **Global** scope
- **Local** scope

Variables declared outside of a function have *global* scope. All expressions in the same script, inside or outside of functions, can access global variables. Variables defined inside a function have *local* scope. Local variables are only accessible from expressions inside the same function where they are defined.

In the following code section, `carName` is a global variable. It can be used inside and outside of the `myFunction` function:

```
var carName = "Suzuki";
// Code here can use carName
function myFunction() {
    // Code here can use carName
}
```

Here `carName` is a local variable. It can only be used inside `myFunction`, where it was defined:

```
// Code here *cannot* use carName
function myFunction() {
    var carName = "Suzuki";
    // Code here can use carName
}
```

If you assign a value to a variable that has not been declared, it will automatically be defined as a global variable. In the following code example, `carName` is a global variable even though the value is assigned inside a function:

```
// Code here can use carName
function myFunction() {
    carName = "Suzuki";
    // Code here can use carName
}
```

It is not recommended to create global variables unless you intend to, since they can conflict (override) other variables having the same name in the global environment.

---

[16]https://en.wikipedia.org/wiki/Scope_(computer_science)

3.10 Flow control

3.10.1 What is flow control?

By default, all expressions in our script are executed in the given order, top to bottom. This behavior can be modified using **flow control** expressions. There are two types of flow control expressions: conditionals and loops.

- **Conditionals** are used to condition code execution based on different criteria.
- **Loops** are used to execute a block of code a number of times.

3.10.2 Conditionals

The `if` **conditional** is the most commonly used conditional. It is used to specify that a block of JavaScript code will be executed if a condition is true. An `if` conditional is defined as follows:

```
if(condition) {
    // Code to be executed if the condition is true
}
```

For example, the following conditional sets `greeting` to "Good day" if the value of `hour` is less than 18:

```
if(hour < 18) {
    greeting = "Good day";
}
```

The `else` statement can be added to specify an *alternative* block of code, to be executed if the condition is false:

```
if(condition) {
    // Code to be executed if the condition is true
} else {
    // Code to be executed if the condition is false
}
```

For example, the following conditional sets `greeting` to "Good day" if `hour` is less than 18, or to "Good evening" *otherwise*:

```
if(hour < 18) {
    greeting = "Good day";
} else {
    greeting = "Good evening";
}
```

Decision trees of any complexity can be created by combining numerous `if else` expressions. For example, the following set of conditionals defines that if time is less than 10:00, set `greeting` to "Good morning", if not, but time is less than 18:00, set `greeting` to "Good day", otherwise set `greeting` to "Good evening":

```
if(hour < 10) {
    greeting = "Good morning";
} else {
    if(hour < 18) {
        greeting = "Good day";
    } else {
        greeting = "Good evening";
    }
}
```

This type of code may be used to create a customized greeting on a website, for instance. As another example, think of the way you can **toggle** layers on and off in a web map by using an `if` statement to see whether the layer is currently visible. If visible is true, hide the layer, and vice versa.

Conditionals can also be used in map symbology, as we will see later on (Section 8.4). Consider the following example of a function that determines color based on an attribute. The function uses conditionals to return a color based on the "party" attribute: `"red"` for Republican, `"blue"` for Democrat, and `"grey"` for anything else:

```
function party_color(p) {
    if(p === "Republican") return "red"; else
    if(p === "Democrat") return "blue"; else
    return "grey";
}
```

Note that in the above example, the brackets (`{}`) around conditional code blocks are omitted to make the code shorter, which is legal when code blocks consist of a single expression[17].

___

- Define the `party_color` function, by copying the above code and pasting it in the console.
- Try running the `party_color` function three times, each time with a different argument, to see how you can get all three possible returned values: `"red"`, `"blue"`, or `"grey"`.

___

[17]For more information on conditionals, see the *Making decisions in your code—conditionals* article (`https://developer.mozilla.org/en-US/docs/Learn/JavaScript/Building_blocks/conditionals`) by Mozilla.

3.10.3 Loops

3.10.3.1 Standard `for` loop syntax

Loops are used to execute a piece of code a number of times. JavaScript supports several types of loops. The most useful kind of loop within the scope of this book is the `for` loop.

The **standard `for`** loop has the following syntax:

```
for(statement 1; statement 2; statement 3) {
    // Code block to be executed
}
```

where:

- **Statement 1** is executed once, before the loop starts.
- **Statement 2** defines the condition to keep running the loop.
- **Statement 3** is executed in each "round," after the code block is executed.

For example:

```
var text = "";
for(var i = 0; i < 5; i++) {
    text += "The number is " + i + "<br>";
}
```

Let's try to locate the `for` loop components in the above example:

- **Statement 1** sets a variable before the loop starts (`var i=0`).
- **Statement 2** defines the condition for the loop to keep running (`i<5` must be `true`, i.e., i must be less than five).
- **Statement 3** increases the value of i (`i++`) each time the code block in the loop has been executed, using the increment operator `++` (Section 3.6.2.2).

As a result, the code block is executed five times, with i values of 0, 1, 2, 3, and 4. A loop is thus an alternative to repetitive code. For example, the above `for` loop is an alternative to the following code section, which gives exactly the same result *without* using a loop:

```
var text = "";
text += "The number is " + 0 + "<br>";
text += "The number is " + 1 + "<br>";
text += "The number is " + 2 + "<br>";
text += "The number is " + 3 + "<br>";
text += "The number is " + 4 + "<br>";
```

- Execute the `for` loop from the above code section, then print the value of the `text` variable to see the effect of the loop.

Here is another example of a `for` loop. In this example, the code will be executed 1000 times since i gets the values of 0, 1, 2, etc., up to 999.

```
for(var i = 0; i < 1000; i++) {
    // Code here will run 1000 times
}
```

3.10.3.2   Iterative `for` loop syntax

In addition to the standard `for` loop syntax shown above, there is a special syntax to iterate over elements of arrays or objects. In this **iterative** syntax, instead having three statements, we define the loop with just one statement:

```
for(var i in object) {
    // Code block to be executed
}
```

where i is a variable name (of our choice) and `object` is the object which we iterate over. In each iteration, i is set to another `object` element name. In case `object` is an array, i gets the array index values: "0", "1", "2", and so on. In case `object` is an object, i gets the property names. For example:

```
var obj = {a: 12, b: 13, c: 14};
for(var i in obj) {
    console.log(i + " " + obj[i]);
}
```

This loop runs three times, once for each property of the object `obj`. Each time, the i variable gets the next property name of `obj`. Inside the code block, the current property name (i) and property value (`obj[i]`) are being printed, so that the following output appears in the console:

```
a 12
b 13
c 14
```

For the next `for` loop example, we define an object named `directory`:

```
var directory = {
    musicians: [
        {firstname: "Chuck", lastname: "Berry"},
        {firstname: "Ray", lastname: "Charles"},
        {firstname: "Buddy", lastname: "Holly"}
    ]
};
```

The `directory` object has just one property, named `musicians`, which is an array. Each element in the `directory.musicians` array is an object, with `firstname` and `lastname` properties. Using a `for` loop we can go over the `musicians` array, printing the full name of each musician:

```
for(var i in directory.musicians) {
    var musician = directory.musicians[i];
    console.log(musician.firstname + " " + musician.lastname);
}
```

This time, since `directory.musicians` is an array, `i` gets the array indices "0", "1", and "2". These indices are used to select the current item in the array, with `directory.musicians[i]`. As a result, the following output is printed in the console[18]:

```
Chuck Berry
Ray Charles
Buddy Holly
```

3.11 JavaScript Object Notation (JSON)

3.11.1 JSON

JavaScript Object Notation (JSON)[19] (Bassett, 2015) is a data format closely related to JavaScript objects. It is a **plain text** format, which means that a JSON instance is practically a character string, which can be saved in a plain text file (usually with the `.json` file extension), in a database, or in computer memory[20].

JSON and JavaScript objects are very similar and easily interchangeable. For that reason, JSON is the most commonly used format for exchanging data between the server and the client in web applications. The principal difference between JSON and JavaScript objects is as follows:

- A JavaScript **object** is a data type in the JavaScript environment. A JavaScript object does not make sense outside of the JavaScript environment.
- A **JSON** instance is a plain text string, formatted in a certain way according to the JSON standard[21]. A JSON string is thus not limited to the JavaScript environment. It can exist in another programming language, or simply stored inside a plain text file or a database.

For example, we already saw that a JavaScript object can be created using an expression such as the following one:

```
var obj = {a: 1, b: 2};
```

---

[18]For more information on loops, see the *Looping code* article (https://developer.mozilla.org/en-US/docs/Learn/JavaScript/Building_blocks/Looping_code) by Mozilla.

[19]https://en.wikipedia.org/wiki/JSON

[20]Other well-known plain text data formats are, for instance, *Comma-Separated Values (CSV)* (https://en.wikipedia.org/wiki/Comma-separated_values) and *Extensible Markup Language (XML)* (https://en.wikipedia.org/wiki/XML).

[21]https://json.org/

The corresponding JSON string can be defined, and stored in a variable, as follows:

```
var json = '{"a": 1, "b": 2}';
```

A notable difference between how the above two variables are defined is that in a JSON string the property names are enclosed in quotes: `"a"` and `"b"` (we will immediately explain why). The JSON standard requires double quotes (`"`), so we need to enclose the entire string with single quotes (`'`).

To make a JSON string useful within the JavaScript environment, the JSON string can be **parsed** to produce an object. The parsing process, **JSON→object**, is done with the `JSON.parse` function. The `JSON.parse` function is used to convert a JSON string (e.g., coming from a server) to a JavaScript object:

```
JSON.parse(json);   // Returns an object
```

As we just saw, a JSON string should have quoted property names, as in `'{"key": "value"}'` and NOT `'{key: "value"}'`. The purpose of this convention is to avoid parsing errors in cases when a key name is a **reserved** JavaScript keyword. For example, the following expression gives an error, because `while` is a reserved keyword in JavaScript:

```
JSON.parse('{while: 5}');
```

The following expression, however, will work fine, because the word `while` has been enclosed in quotes:

```
JSON.parse('{"while": 5}');
```

The opposite conversion, **object→JSON**, is done with `JSON.stringify`:

```
JSON.stringify(obj);   // Returns a JSON string
```

The `JSON.stringify` function is commonly used when sending an object from the JavaScript environment to a server. Since, like we said, JavaScript objects cannot exist outside of the JavaScript environment, we need to convert the object to a string with `JSON.stringify` before the data are sent elsewhere. For example, we are going to use `JSON.stringify` in Chapter 13 when sending spatial layers to permanent storage in a database (Section 13.6).

Finally, keep in mind that JSON is limited in that it cannot store **undefined** values or functions. JSON *can* store all of the other JavaScript data types (Section 3.6.1)[22]:

- Strings
- Numbers
- Booleans
- `null`
- Arrays
- Objects

---

[22]For more details on the difference between a JavaScript object and a JSON string, check out the *StackOverflow question* on this matter (https://stackoverflow.com/questions/6489783/whats-the-difference-between-javascript-object-and-json-object).

3.11.2 GeoJSON

GeoJSON[23] is a spatial vector layer format based on JSON. Since GeoJSON is a special case of JSON, it can be easily processed in the JavaScript environment using the same methods as any other JSON string, such as `JSON.parse` and `JSON.stringify` (Section 3.11.1). For this reason, GeoJSON is the most common data format for exchanging spatial (vector) data on the web.

Here is an example of a GeoJSON string:

```
{
  "type": "Feature",
  "geometry": {
    "type": "Point",
    "coordinates": [125.6, 10.1]
  },
  "properties": {
    "name": "Dinagat Islands"
  }
}
```

This particular GeoJSON string represents a point layer with one attribute called **name**. The layer has just one feature, a point at coordinates `[125.6, 10.1]`. Don't worry about the details of the GeoJSON format at this stage. We will cover the syntax of the GeoJSON format in Chapter 7.

A variable containing the above GeoJSON string can be created in a JavaScript environment with the following expression:

```
var pnt = '{' +
  '"type": "Feature",' +
  '"geometry": {' +
    '"type": "Point",' +
    '"coordinates": [125.6, 10.1]' +
  '},' +
  '"properties": {' +
    '"name": "Dinagat Islands"' +
  '}' +
'}';
```

Note that the string is defined piece by piece, concatenating with the + operator, to make the code more readable. We could likewise define **pnt** in a single line, using a long string and without needing **+**.

GeoJSON can be parsed with `JSON.parse` just like any other JSON:

```
pnt = JSON.parse(pnt);   // Returns an object
```

---

[23]https://en.wikipedia.org/wiki/GeoJSON

Now that `pnt` is a JavaScript object, we can access its contents using the dot (or bracket) notation just like with any other object:

```
pnt.type;                      // Returns "Feature"
pnt.geometry.coordinates;      // Returns [125.6, 10.1]
pnt.geometry.coordinates[0];   // Returns 125.6
pnt.properties;                // Returns {name: "Dinagat Islands"}
```

Going back to a GeoJSON string is done with `JSON.stringify`:

```
JSON.stringify(pnt);   // Returns a string
```

The above expression gives the following result[24]:

```
{"type":"Feature","geometry":{"type":"Point","coordinates":[...]}}
```

Note that `JSON.stringify` takes further arguments to control the formatting of the resulting string. For example, `JSON.stringify(pnt, null, 4)` will create the following indented, multi-line string—much like the one we began with. This is much easier to read than the default one-line result shown above:

```
{
    "type": "Feature",
    "geometry": {
        "type": "Point",
        "coordinates": [
            125.6,
            10.1
        ]
    },
    "properties": {
        "name": "Dinagat Islands"
    }
}
```

3.12 Exercise

- Write a short JavaScript code, as follows.
- The code starts with defining two arrays:
  - The *first* array represents a single two-dimensional coordinate A, such as: `[100, 200]`.
  - The *second* array is an array of coordinate pairs B, such as: `[[0, 3], [0, 4], [1, 5], [2, 5]]`.

---

[24]The last part of the string is omitted and replaced with `[...]` to fit on the page.

- The code you write should create a *new* array, where the individual coordinate A is attached to each of the coordinates in B, so that we get an array of coordinate pairs, such as: `[[[100, 200], [0, 3]], [[100, 200], [0, 4]], [[100, 200], [1, 5]], [[100, 200], [2, 5]]]`[25].
- Hint: create an empty array `result`, then run a loop that goes over the second array B, each time "pushing" (Section 3.8.2) the A element plus the current B element into `result`.

---

[25]The resulting array can be useful for drawing line segments between a single point A and a second set of points B (Figure 11.8).

4

JavaScript Interactivity

4.1   Introduction

In Chapters 1–2, we mentioned that JavaScript is primarily used to control the interactive behavior of web pages. However, in Chapter 3 we introduced the JavaScript language "in isolation": none of the JavaScript expressions we executed in the console had any effect on the contents of the web page where they were run. Indeed, how can we link JavaScript code with page contents, to interactively modify the web page? This is precisely the subject of the current chapter—employing JavaScript to query and modify web page content. As we will see, the mechanism which makes the link possible is the **Document Object Model (DOM)**.

4.2   The Document Object Model (DOM)

When a web browser loads an HTML document, it displays (renders) the contents of that document on the screen, possibly styled according to CSS styling rules. But that's not all the web browser does with the tags, attributes, and text contents of the HTML document. The browser also creates and memorizes a "model" of that page's HTML (Figure 4.1). In other words, the browser remembers the HTML tags, their attributes, and the order in which they appear in the file. This representation of the page is called the **Document Object Model (DOM)**[1].

The DOM is sometimes referred to as an **Application Programming Interface (API)**[2] for accessing HTML documents with JavaScript. An API is a general term for methods of communication between software components. The DOM can be considered an API, since it bridges between JavaScript code and page contents displayed in the browser. The DOM basically provides the information and tools necessary to navigate through, or make changes or additions, to the HTML on the page. The DOM itself isn't actually JavaScript—it's a standard from the World Wide Web Consortium (W3C)[3] that most browser manufacturers have adopted and added to their browsers.

---

[1]https://en.wikipedia.org/wiki/Document_Object_Model
[2]https://en.wikipedia.org/wiki/Application_programming_interface
[3]https://www.w3.org/DOM/

```
<!DOCTYPE html>
<html>
    <head>
        <title>A minimal web page</title>
    </head>
    <body>
        <h1>This is the main heading</h1>
        <p>This is a paragraph.</p>
    </body>
</html>
```

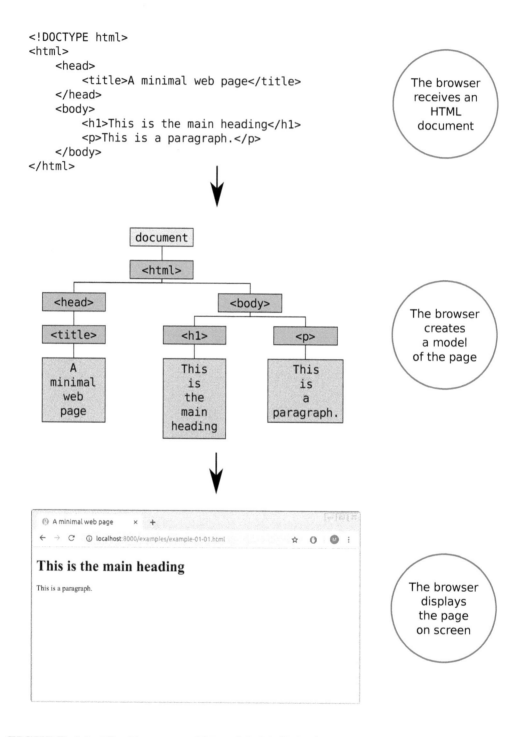

FIGURE 4.1: The Document Object Model (DOM) is a representation of an HTML document, constructed by the browser on page load

4.3 Accessing and modifying elements

4.3.1 Overview

All of the properties and methods of the DOM, available for manipulating web pages, are organized into objects which are accessible through JavaScript code. Two of the most important objects are:

- The `window` object, which represents the browser **window**, i.e., the global environment
- The `document` object, which represents the **document** itself[4]

Using the properties and methods of the `window` and `document` objects, we are able to access document element contents or their display settings, and also dynamically adjust them. For example, the `document` object has a property named `title`, referring to the text within the `<title>` element—the page title (Section 1.6.2.1). We can access `document.title`, to get a string with page title, or modify page title, by assigning a new string to `document.title`.

- Open the browser and browse to any web page you like.
- Open the JavaScript console (Section 3.4) and try typing `document.title` and `window.location.href`.
- Check the type of returned values, using the `typeof` operator (Section 3.6.4).
- Try assigning a new string into `document.title`. The new title should appear on top of the browser window!
- Go to any page (other than `https://www.google.com`) and type `window.location="https://www.google.com"` into the console. What do you think has happened?

4.3.2 Accessing elements

The `document.title` property was just an example, referring to the specific `<title>` element. We need a more general method if we want to be able to locate, and modify, *any* element in our document. The `document` object indeed contains several methods for finding elements in the DOM. These methods are called DOM **selectors**, or DOM queries. The following two expressions are examples of DOM selectors.

```
document.getElementById("firstParagraph");
document.getElementsByClassName("important");
```

Both expressions employ methods of the `document` object:

- The *first* selector uses the `.getElementById` method to select an individual element with a specific ID, `id="firstParagraph"`.
- The *second* selector uses the `.getElementsByClassName` method to select all elements with a specific class, `class="important"`.

---

[4]The `document` object, like all other global objects, is in fact a property of the `window` object. That is, `document` is a shortcut for `window.document`.

4.3.3   Modifying elements

The result of a DOM query is a **reference** to an element, or set of elements, in the DOM. That reference can be used to update the contents or behavior of the element(s). For example, the `innerHTML` property of a DOM element refers to the HTML code of that element. The following expression uses the `innerHTML` property to modify the HTML content of the element which has `id="firstParagraph"`, replacing any previous HTML code inside that element with `<b>Hello!</b>`.

```
document.getElementById("firstParagraph").innerHTML = "<b>Hello!</b>";
```

It is important to understand that the HTML **source code** and the **DOM** are two separate entities. While the HTML source code sent to the browser is constant, the DOM, which is initially constructed from the HTML source code (Figure 4.1), can be dynamically altered using JavaScript code. As long as no JavaScript code that modifies the DOM is being run, the DOM and the HTML source code are identical. However, when JavaScript code *does* modify the DOM, the DOM changes and the displayed content in the browser window changes accordingly.

The current DOM state can be examined, for example, using the **Elements** tab of the **developer tools** (in Chrome). The HTML source code can be shown with **Ctrl+U** (in Chrome), as we have already discussed in Section 1.3. Again, the source code remains exactly the same as initially sent from the server, no matter what happens on the page, while the DOM may be modified by JavaScript code and thus may change. This means that the HTML source code does not necessarily match the DOM once any JavaScript code was executed.

- While running the examples in this chapter, you can compare the HTML source code and the DOM to see how the DOM changes in response to executed JavaScript code, while the HTML source code remains the same as initially loaded.

4.3.4   Event listeners

Sometimes we want to change the DOM at once, for example on page load. In other cases, however, we want to change the DOM *dynamically*, in response to user actions on the page, such as clicking on a button. This is where the DOM **events** come into play. Each and every thing that happens to a web page is called an event. Web browsers are programmed to recognize various events, including user actions such as:

- Mouse movement and clicks
- Pressing on the keyboard
- Re-sizing the browser window

An event represents the precise moment when something happens inside the browser. This is sometimes referred to as the event being *fired*. There are different types of events, depending on the type of action taking place. For example, when you click a mouse, at the precise moment you release the mouse button the web browser signals that a `"click"` event has just occurred. In fact, web browsers fire several separate events whenever you click the mouse button. First, as soon as you press the mouse button, the `"mousedown"` event fires; then, when you let go of the button, the `"mouseup"` event fires; and finally, the `"click"` event fires.

To make your web page interactive, you need to write code that runs and does something useful in response to the appropriate type of event occurring on the appropriate element(s). This type of binded code is known as an **event listener**, or an event handler. For example, we may wish to set an event listener which responds to user click on an interactive map by adding a marker in the clicked location (Section 11.2.2). In such case, we need to bind an event listener:

- to the interactive map object,
- with a function for adding a marker,
- which is executed in response to a mouse `"click"` event.

A mouse click is just one example, out of many predefined event types that the browser can detect. Table 4.1 lists some commonly used event types.

TABLE 4.1: Commonly used browser event types

Type	Event	Description
Mouse	`"click"`	Click
	`"dblclick"`	Double-click
	`"mousedown"`	Mouse button pressed
	`"mouseup"`	Mouse button released
	`"mouseover"`	Mouse cursor moves into element
	`"mouseout"`	Mouse cursor leaves element
	`"mousemove"`	Mouse moved
	`"drag"`	Element being dragged
Keyboard	`"input"`	Value changed in `<input>` or `<textarea>`
	`"keydown"`	Key pressed
	`"keypress"`	Key pressed (character keys only)
	`"keyup"`	Key released
Focus and Blur	`"focus"`	Element gains focus (e.g., typing inside `<input>`)
	`"blur"`	Element loses focus
Forms	`"submit"`	Form submitted
	`"change"`	Form changed
Document/Window	`"load"`	Page finished loading
	`"unload"`	Page unloading (new page requested)
	`"error"`	JavaScript error encountered
	`"resize"`	Browser window resized
	`"scroll"`	User scrolls the page

In this book, we will mostly use the `"click"`, `"mouseover"`, `"mouseout"`, and `"drag"` events, all from the Mouse events category, as well as `"change"` from the Forms category (Table 4.1). However, it is important to be aware of the other possibilities, such as events related to window resize or scrolling, or keyboard key presses.

4.3.5   Hello example

The next two examples demonstrate the idea of event listeners using **plain** JavaScript. We will not go into details in terms of syntax, though, because starting from Section 4.4 we will be using an easier way of doing the same things with the **jQuery** JavaScript library.

Consider the following HTML code of `example-04-01.html`:

```html
<!DOCTYPE html>
<html>
    <head>
        <title>Hello JavaScript</title>
    </head>
    <body>
        <h2>JavaScript</h2>
        <p id="demo">What can JavaScript do?</p>
        <input type="button" value="Click Me!" id="change_text">
        <script>
            function hello() {
                document
                    .getElementById("demo")
                    .innerHTML =
                        "JavaScript can change page contents!";
            }
            document
                .getElementById("change_text")
                .addEventListener("click", hello);
        </script>
    </body>
</html>
```

In this example, we have a web page with an **<h2>** heading (*without* an ID), as well as two other elements *with* an ID:

- A **<p>** element with **id="demo"**
- An **<input>** button element with **id="change_text"**

In the end of the **<body>**, we have a **<script>** element containing JavaScript code (Section 1.6.3). Since the **<script>** element is in the end of the **<body>**, it will be executed by the browser *after* the HTML code is loaded and processed. Let's look at the last line in the JavaScript code:

```
document.getElementById("change_text").addEventListener("click", hello);
```

The above expression does several things:

- **Selects** the element which has **id="change_text"** (the *button*), using the **document.getElementById** method (Section 4.3.2)
- **Binds** an event listener to it, using the **.addEventListener** method
- The event listener specifies that when the element is clicked (i.e., the **"click"** event is fired), the **hello** function will be executed

What does the `hello` function do? According to its definition at the beginning of the `<script>`, we see that it has no parameters and just one expression:

```
function hello() {
    document
        .getElementById("demo")
        .innerHTML = "JavaScript can change page contents!";
}
```

What the `hello` function does is:

- **Selects** the element with `id="demo"` (the *paragraph*), again using the `document.getElementById` method
- **Replaces** its HTML contents with `"JavaScript can change page contents!"`, by assigning the new contents into the `innerHTML` property (Section 4.3.3)

Note that in this example the expression inside the `hello` function is incidentally split into several (three) lines, to fit on the page. However, new-line symbols and repeated spaces are ignored by the JavaScript interpreter. You can imagine the characters being merged back into a long single-line expression, then executed. In other words, the computer still sees a single expression here, where we assign a string into the `.innerHTML` property of an element selected using `document.getElementById`.

The way that `example-04-01.html` appears in the browser is shown in Figure 4.2.

FIGURE 4.2: Screenshot of `example-04-01.html`

- Open `example-04-01.html` in the browser, then open the Elements tab in Developer Tools.
- Click the button that says "Click me!" (Figure 4.2), and observe how the paragraph contents are being modified.
- Note how the affected part of the DOM is momentarily highlighted each time the button is pressed.

4.3.6   Poles example

The second example `example-04-02.html` is slightly more complex, but the principle is exactly the same as in `example-04-01.html`:

```
<!DOCTYPE html>
<html>
    <head>
        <title>Earth poles</title>
    </head>
    <body>
        <h2>Earth Poles Viewer</h2>
        <img id="myImage" src="images/north.svg"><br>
        <input type="button" value="North Pole" id="north">
        <input type="button" value="South Pole" id="south">
        <script>
            function showNorth() {
                document
                    .getElementById("myImage")
                    .src = "images/north.svg";
            }
            function showSouth() {
                document
                    .getElementById("myImage")
                    .src = "images/south.svg";
            }
            document
                .getElementById("north")
                .addEventListener("click", showNorth);
            document
                .getElementById("south")
                .addEventListener("click", showSouth);
        </script>
    </body>
</html>
```

In this example, we have two buttons, and we add two event listeners: one for the North pole button and one for the South pole button (Figure 4.3). Both event listeners change the `src` attribute of the `<img>` element on the web page. The North pole button sets `src='images/north.svg'`, while the South pole button sets `src='images/south.svg'`. The resulting effect is that the displayed globe is switched from viewing the North pole or the South pole[5].

Note that the images loaded in this example—`north.svg` and `south.svg`—are in a format called **Scalable Vector Graphics (SVG)**[6], having the `.svg` file extension. SVG is an open standard *vector* image format. It is well supported by web browsers and commonly used to display vector graphics on web pages.

The result of `example-04-02.html` is shown in Figure 4.3.

FIGURE 4.3: Screenshot of `example-04-02.html`

4.4 What is jQuery?

So far we introduced the DOM (Section 4.2), and saw how it can be used to (dynamically) modify page contents (Section 4.3). In the remainder of this chapter, we will learn about the jQuery library which simplifies these (and other) types of common tasks.

A JavaScript **library**[7] is a collection of JavaScript code, which allows for easier development of JavaScript-based applications. There are a lot of JavaScript libraries that simplify common tasks (e.g., DOM manipulation) or specialized tasks (e.g., web mapping), to make life easier for web developers. Often, you will be working with a library that is already written, instead of "reinventing the wheel" and writing your own JavaScript code for every single task.

---

[5]We will come back to the subject of specifying file paths in Section 5.5.
[6]https://en.wikipedia.org/wiki/Scalable_Vector_Graphics
[7]https://en.wikipedia.org/wiki/JavaScript_library

jQuery[8] is a JavaScript library that simplifies tasks related to interaction with the DOM, such as selecting elements and binding event listeners—like we did in the last two examples. Since this type of task is very common in JavaScript code, jQuery is currently the most widely used[9] JavaScript library, by a large margin.

The main functionality of jQuery consists of:

- **Finding** elements using CSS-style selectors
- **Doing** something with those elements, using jQuery methods

In the following Sections 4.5–4.10, we are going to introduce these two concepts while writing alternative implementations of the "Hello" (Sections 4.3.5) and "Poles" (Section 4.3.6) examples. In the new versions of the examples (Sections 4.9–4.10), we will be using jQuery, instead of plain JavaScript, for selecting elements and for binding event listeners.

In addition to its main functionality, jQuery has functions and methods to simplify other types of tasks in web development. For example, the `$.each` function (Section 4.12) simplifies the task of **iterating** over arrays and objects. In later chapters, we will learn about a technique for loading content into our web page called **Ajax**, which is also simplified using jQuery (Section 7.7).

4.5   Including the jQuery library

4.5.1   Including a library

Before using any object, function, or method from jQuery, the library needs to be **included** in our web page. Practically, this means that the jQuery script is being run on page load, defining jQuery objects, functions, and methods, which you can then use in the subsequent scripts that are executed on that page.

To include the jQuery library—or any other script for that matter—we need to place a `<script>` element referring to that script in our HTML document. Scripts for loading libraries are commonly placed inside the `<head>` of our document. Placing a `<script>` in the `<head>` means that the script is loaded before anything visible (i.e., the `<body>`) is loaded. This can be safer and easier for maintenance, yet with some performance implications. Namely, page load is being halted until the `<scripts>` elements have been processed. Since the jQuery script is very small (~90 kB), there is no noticeable delay for downloading and processing it, so we can safely place it in the `<head>`.

As mentioned in Section 1.6.3, when using the `<script>` element to load an external script file, we use the `src` attribute to specify file location. The location specified by `src` can be either a path to a local file, or a URL of a remote file hosted elsewhere on the web.

4.5.2   Loading a local script

When loading a script from a **local file**, we need to have an actual copy of the file on our server (more on that in Section 5.5). Basically, we need to download the jQuery code file,

---

[8]https://jquery.com/
[9]https://w3techs.com/technologies/overview/javascript_library/all

e.g., from the download section[10] on the official jQuery website, and save it along with our HTML file. In case the jQuery script we downloaded is named `jquery.js`, the first few lines of the document may look as follows:

```
<!DOCTYPE html>
<html>
    <head>
        <script src="jquery.js"></script>
    </head>
...
```

4.5.3 Loading a remote script

When loading a script from a **remote file**, hosted on the web in some location other than our own computer, we need to provide the file URL. A reliable option is to use a **Content Delivery Network (CDN)**[11], such as the one provided by Google[12]. A CDN is a series of servers designed to serve static files very quickly. In case we are loading the jQuery library from Google's CDN, the first few lines of the document may look as follows:

```
<!DOCTYPE html>
<html>
    <head>
        <script src="https://ajax.[...]/jquery.min.js"></script>
    </head>
...
```

The `src` attribute value is truncated with `[...]` to save space. Here is the complete URL that needs to go into the `src` attribute value:

```
https://ajax.googleapis.com/ajax/libs/jquery/3.4.1/jquery.min.js
```

- Browse to the above URL to view the jQuery library contents. You can also download the file by clicking **Save as...** to view it in a text editor instead of the browser.
- You will note that the code is formatted in a strange way, as if the entire content is in a single line, with very few space and new line characters. This makes it hard for us to read the code.
- Can you guess what is the reason for this type of formatting?

Whether we refer to a local file or to a remote file, the jQuery code will be loaded and processed by the browser, which means we can use its functions and methods in other scripts on that page.

---

[10] http://jquery.com/download/
[11] https://en.wikipedia.org/wiki/Content_delivery_network
[12] https://developers.google.com/speed/libraries/

4.6   Selecting elements

As mentioned above, the main functionality of jQuery involves selecting elements and then doing something with them. With jQuery, we usually select elements using CSS-style selectors. To make a selection, we use the $ function defined in the jQuery library. The $ function can be invoked with a single parameter, the **selector**.

Most of the time, we can apply just the three basic CSS selector types (which we covered in Section 2.4) to get the selection we need. Using these selectors, we target elements based on their type, ID, or class. For example:

```
$("a");          // Selects all <a> elements
$("#choices");   // Selects the element with id="choices"
$(".submenu");   // Selects all elements with class="submenu"
```

The result of each of these expressions is a jQuery object, which contains a *reference* to the selected DOM element(s) along with methods for doing something with the selection. For example, the jQuery object has methods for modifying the contents of all selected elements (Sections 4.7) or adding event listeners to them (Section 4.8), as discussed below.

It should be noted that jQuery also lets you use a wide variety of advanced CSS selectors to accurately pinpoint the specific elements we need inside complex HTML documents. We will not be needing such selectors in this book[13].

4.7   Operating on selection

4.7.1   Overview

jQuery objects are associated with numerous methods for acting on the selected elements, from simply replacing HTML (Section 4.7.2), to precisely positioning new HTML in relation to a selected element (Sections 4.7.4–4.7.5), to completely removing elements and content from the page. Table 4.2 lists some of the most useful jQuery methods[14] for operating on selected elements, which we cover in the following Sections 4.7.2–4.7.7.

TABLE 4.2:   jQuery methods for operating on selection

Type	Method	Description
Getting/Changing content	.html	Get or set content as HTML
	.text	Get or set content as text
Adding content	.append	Add content before closing tag
	.prepend	Add content after opening tag

---

[13]To get an impression of the various advanced selector types in jQuery, check out the interactive demonstration (https://www.w3schools.com/jquery/trysel.asp) by W3Schools, where the first few examples show basic selectors while the other examples show advanced ones.

[14]There are many more jQuery methods that you can use. For other examples see the *Working with Selections* (https://learn.jquery.com/using-jquery-core/working-with-selections/) article by jQuery

Type	Method	Description
Attributes and values	.attr	Get or set attribute value
	.val	Get or set value of input element

The following small web page (example-04-03.html) will be used to demonstrate selecting elements and acting on selection, using jQuery.

```html
<!DOCTYPE html>
<html>
    <head>
        <title>jQuery operating on selection</title>
        <script src="js/jquery.js"></script>
    </head>
    <body>
        <h1 id="header">List</h1>
        <h2>Buy groceries</h2>
        <ul>
            <li id="one" class="hot"><i>fresh</i> figs</li>
            <li id="two" class="hot">pine nuts</li>
            <li id="three" class="hot">honey</li>
            <li id="four">balsamic vinegar</li>
        </ul>
        <h2>Comments</h2>
        <input type="text" id="test3" value="Mickey Mouse">
    </body>
</html>
```

Figure 4.4 shows how example-04-03.html appears in the browser.

You can open example-04-03.html in the browser and run the expressions shown in Sections 4.7.2–4.7.7 (below) in the console, to see their immediate effect on page content.

4.7.2   .html

The .html method can be used to read the current HTML code inside an element, or to replace it with some other HTML code. To retrieve the HTML currently inside the selection, just add .html() after the jQuery selection. For example, you can run the following command in the console where example-04-03.html is loaded:

```
$("#one").html();
```

This should give the following output:

```
"<i>fresh</i> figs"
```

In this example, we used a selector to locate the element with id="one", then used the .html method on the selection to get its HTML contents. Keep in mind that if the selection

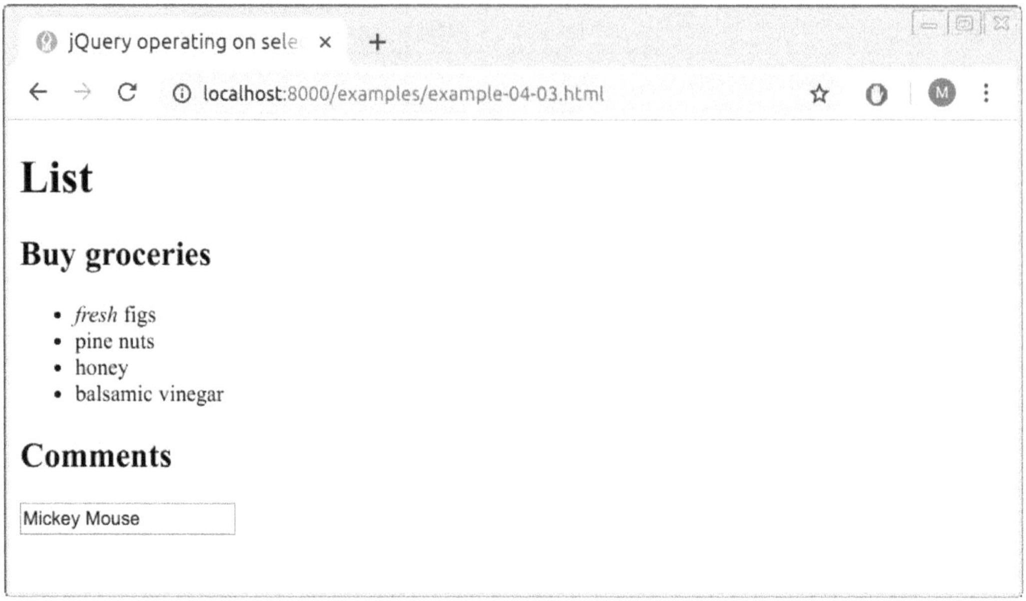

FIGURE 4.4: Screenshot of `example-04-03.html`

contains more than one element, only the contents of the *first* element is returned by the `.html` method. For example, the following expression returns the same value as the previous one, because the first `<li>` element on the page is also the one having `id="one"`:

```
$("li").html();
```

If you supply a string as an argument to `.html`, you *replace* the current HTML contents inside the selection. This can be seen as the jQuery alternative for assignment into the `.innerHTML` property using plain JavaScript (Section 4.3.3):

```
$("#one").html("<i><b>Not very</b> fresh</i> figs");
```

In case the selection contains more than one element, the HTML contents of *all* elements is set to the new value. For example, the following expression will replace the contents of all list items on the page:

```
$("li").html("<i><b>Not very</b> fresh</i> figs");
```

- Open `example-04-03.html` and run the above expression for changing the contents of the first `<li>` element in the console.
- Compare the HTML source code (**Ctrl+U**) and the DOM (**Elements** tab in the **developer tools**).
- Which one reflects the above change and why? (Hint: see Section 4.3.3.)

4.7.3 `.text`

The `.text` method works like `.html`, but it does not accept HTML tags. It is therefore useful when you want to replace the text within an element. For example:

```
$("#two").text("pineapple");
```

- What do you think will happen if you pass text that contains HTML tags–such as `<b>pineapple</b>`–to the `.text` method?
- Try it out in the console to check your answer.

4.7.4 `.append`

The `.append` and `.prepend` methods add new content *inside* the selected elements. The `.append` method adds HTML as the **last child** element of the selected element(s). For example, say you select a `<div>` element, but instead of replacing the contents of the `<div>`, you just want to add some content *before* the closing `</div>` tag. The `.append` method, when applied to the `<div>`, does just that.

The `.append` method is a great way to add an item to the end of a bulleted (`<ul>`) or numbered (`<ol>`) list. For example, the following expression adds a new list item at the end of our `<ul>` element in **example-04-03.html**:

```
$("ul").append("<li>ice cream</li>");
```

An example of a web page that uses `.append` for adding new list items will be given in Section 4.13.

4.7.5 `.prepend`

The `.prepend` method is like `.append`, but adds HTML content as the **first child** of the selected element(s), that is, directly *after* the opening tag. For example, we can use `.prepend` to add a new item at the beginning, rather than the end, of our list:

```
$("ul").prepend("<li>bread</li>");
```

4.7.6 `.attr`

The `.attr` method can get or set the value of a specified **attribute**. To get the value of an attribute of the *first* element in our selection, we specify the name of the attribute in the parentheses. For example, the following expression gets the `class` attribute value of the element that has `id="one"`:

```
$("#one").attr("class");   // Returns "hot"
```

To modify the value of an attribute for *all* elements in our selection, we specify both the attribute name and the new value. For example, the following expression sets the `class` attribute of all `<li>` elements to `"cold"`:

```
$("li").attr("class", "cold");
```

4.7.7   `.val`

The `.val` method gets or sets the current *value* of input elements, such as `<input>` or `<select>` (Section 1.6.12). Note that `.attr("value")` (Section 4.7.6) is not the same as `.val()`. The former method gives the attribute value in the DOM, such as the initial `value` set in HTML code. The latter gives the real-time value set by the user, such as currently entered text in a text box, which is not reflected in the DOM.

For example, the web page we are experimenting with (`example-04-03.html`) contains the following `<input>` element:

```
<input type="text" id="test3" value="Mickey Mouse">
```

The following expression gets the current value of that element, which is equal to `"Mickey Mouse"` unless the user interacted with the text area and typed something else:

```
$("#test3").val();   // Returns "Mickey Mouse"
```

If you modify the contents of the text input area, and run the above expression again, you will see the new, *currently* entered text. The `.attr("value")` method, on the other hand, will still show the original value (`"Mickey Mouse"`).

You may already have guessed that passing an argument to `.val` can be used to modify the current input value. The following expression sets the input value to `"Donald Duck"`, replacing the previous value of `"Mickey Mouse"`, or whatever else that was typed into the text area:

```
$("#test3").val("Donald Duck");
```

We will get back to another practical example using `.val` and `<input>` elements in Section 4.14.

4.8   Binding event listeners

In addition to querying and modifying contents, we can also bind **event listeners** to the elements in our selection. At the beginning of this chapter, we binded event listeners using plain JavaScript (Sections 4.3.4–4.3.6). In this section, we will do the same thing (in a simpler way) using jQuery.

jQuery-selection objects contain the `.on` method, which can be used to add event listeners to all elements in the respective selection. Similarly to the `.addEventListener` method in plain JavaScript which we saw above (Section 4.3.5), the jQuery `.on` method accepts two arguments:

- A specification of the event **type(s)** (Table 4.1) that are being listened to, such as `"click"`
- A **function**, which is executed each time the event fires

For example, to bind a `"click"` event listener to all paragraphs on a page, you can use the following expression:

```
$("p").on("click", myFunction);
```

where `myFunction` is a function that defines *what* should happen each time the user clicks on a `<p>` element.

The function we pass to the event listener does not need to be a predefined named function, such as `myFunction`. You can also use an anonymous function (Section 3.8). For example, here is another version of the above event listener definition, this time using an anonymous function:

```
$("p").on("click", function() {
  // Code goes here
});
```

It is often convenient to add multiple event listeners to the same element selection, in order to trigger different responses for different events. In our next example (`example-04-04.html`), two event listeners[15] are binded to the `id="p1"` paragraph. The first event listener responds to the `"mouseover"` event (mouse cursor entering the element), printing `"You entered p1!"` in the console. The second event listener responds to the `"mouseout"` event (mouse cursor leaving the element), printing `"You left p1!"` in the console.

```
$("#p1")
    .on("mouseover", function() { console.log("You entered p1!"); })
    .on("mouseout", function() { console.log("You left p1!"); });
```

As a result, the phrases the `"You entered p1!"` or `"You left p1!"` are interchangeably printed in the console whenever the user moves the mouse into the paragraph or out

---

[15]It is possible to bind two (or more) event listeners to the same selection in the *same expression*, thanks to the fact that the `.on` method returns the original selection.

of the paragraph. The small web page implementing the above pair of event listeners (`example-04-04.html`) is shown in Figure 4.5.

FIGURE 4.5: Screenshot of `example-04-04.html`

- Open `example-04-04.html` in the browser, and open the JavaScript console.
- Move the mouse cursor over the paragraph and check out the messages being printed in the console (Figure 4.5).
- Try modifying the source code of this example so that the messages are displayed on the web page itself, instead of the console. (Hint: add another paragraph for the messages in the HTML code, and use the `.text` method to update paragraph contents in response to the events.)

At this stage you may note our code starts to have a lot of nested brackets of both types, (and {. This is typical of JavaScript code, and a common source of errors while learning the language. Make sure you keep track of opening and closing brackets. Most plain text editors, such as Notepad++, automatically highlight the matching opening/closing bracket when placing the cursor on the other one. This can help with code structuring and avoiding errors due to unmatched brackets.

4.9   Hello example

We have now covered everything we need to know to translate the "Hello" (Section 4.3.5) and "Poles" (Section 4.3.6) examples to jQuery syntax. We will start with the "Hello" example where clicking on a button modifies the HTML content of a `<p>` element (Section 4.3.5).

First of all, we need to include the jQuery library (Section 4.5). We add the jQuery library with the following line of code inside the `<head>` element:

```
<script src="js/jquery.js"></script>
```

We are using a *local* file (Section 4.5.2) named `jquery.js`, which is stored in a directory named `js` placed along with our `index.html` file, which is why the file path is specified as `js/jquery.js`. We will elaborate on file structures and file paths for loading content in web pages in Section 5.5.

Next, we replace the following `<script>` contents, which uses plain JavaScript, given in `example-04-01.html` (Section 4.3.5):

```
function hello() {
    document
        .getElementById("demo")
        .innerHTML =
            "JavaScript can change page contents!";
}
document
    .getElementById("change_text")
    .addEventListener("click", hello);
```

with the equivalent jQuery version:

```
function hello() {
    $("#demo").html("JavaScript can change page contents!");
}
$("#change_text").on("click", hello);
```

In the last expression of the jQuery version, the `#change_text` element is selected to bind a `"click"` event listener to it. Whenever the user clicks on the button, the `hello` function is executed. The internal code of the `hello` function also uses a selector, this time accessing the `#demo` element. Once selected, the `.html` function changes the HTML content of the element to `"JavaScript can change page contents!"`.

Note that for a shorter, though perhaps less manageable code, we could use an anonymous function inside the event listener definition:

```
$("#change_text").on("click", function() {
    $("#demo").html("JavaScript can change page contents!");
});
```

Here is the complete code of the modified "Hello" example, as given in `example-04-05.html`:

```
<!DOCTYPE html>
<html>
    <head>
        <title>Hello JavaScript (jQuery)</title>
```

```
        <script src="js/jquery.js"></script>
    </head>
    <body>
        <h2>JavaScript</h2>
        <p id="demo">What can JavaScript do?</p>
        <input type="button" value="Click Me!" id="change_text">
        <script>
            function hello() {
                $("#demo").html("JavaScript can change page contents!");
            }
            $("#change_text").on("click", hello);
        </script>
    </body>
</html>
```

Figure 4.6 shows that the new version is visually (and functionally) the same as the previous one (Figure 4.2); only the underlying code is different, now using jQuery rather than plain JavaScript.

FIGURE 4.6: Screenshot of `example-04-05.html`

4.10 Poles example

Let's modify the "Poles" example to use jQuery too. Again, we need to include the jQuery library in the **<head>** element. Then, we replace the original **<script>** which we saw in Section 4.3.6:

```
function showNorth() {
    document.getElementById("myImage").src = "images/north.svg";
}
function showSouth() {
    document.getElementById("myImage").src = "images/south.svg";
}
document.getElementById("north").addEventListener("click", showNorth);
document.getElementById("south").addEventListener("click", showSouth);
```

with the following alternative version:

```
function showNorth() {
    $("#myImage").attr("src", "images/north.svg");
}
function showSouth() {
    $("#myImage").attr("src", "images/south.svg");
}
$("#north").on("click", showNorth);
$("#south").on("click", showSouth);
```

The concept is similar to the previous example, only that instead of changing the HTML content of an element with the `.html` method, we are changing the `src` attribute with the `.attr` method. Here is the complete code of the modified "Poles" example, as given in `example-04-06.html`:

```
<!DOCTYPE html>
<html>
    <head>
        <script src="js/jquery.js"></script>
    </head>
    <body>
        <h2>Earth Poles Viewer</h2>
        <img id="myImage" src="images/north.svg"><br>
        <input type="button" value="North Pole" id="north">
        <input type="button" value="South Pole" id="south">
        <script>
            function showNorth() {
                $("#myImage").attr("src", "images/north.svg");
            }
            function showSouth() {
                $("#myImage").attr("src", "images/south.svg");
            }
            $("#north").on("click", showNorth);
            $("#south").on("click", showSouth);
        </script>
    </body>
</html>
```

The result is shown in Figure 4.7.

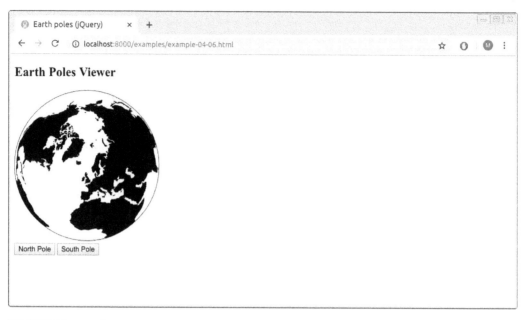

FIGURE 4.7: Screenshot of `example-04-06.html`

4.11   The event object

So far, the functions which we passed to an event listener did not use any information regarding the event itself, other than the fact the event has *happened*. Sometimes, however, we may be interested in functions that have a variable effect, depending on event properties: such as *where* and *when* the event happened.

In fact, every time an event happens, an **event object** is passed to any event listener function responding to the event. We can use the event object to construct functions with variable effects depending on event properties. The event object has methods and properties related to the event that occurred. For example[16]:

- `.type`—Type of event (`"click"`, `"mouseover"`, etc.)
- `.key`—Button or key that was pressed
- `.pageX`, `.pageY`—Mouse position
- `.timeStamp`—Time in milliseconds, from when the document was opened
- `.target`—The element that triggered the event (see Section 8.8.1)

Every function that responds to an event can take the event object as its parameter. That way, we can use the event object properties in the function body, to trigger a specific action according to the properties of the event.

The following small page `example-04-07.html` uses the `.pageX` and `.pageY` properties of the `"mousemove"` event to display up-to-date mouse coordinates every time the mouse moves on screen. Note how the event object–conventionally named `e`, but you can choose another name–is now a parameter of the event listener function. Also note that we are using

---

[16]A list of all standard event object properties can be found in the *HTML DOM Events* (https://www.w3schools.com/jsref/dom_obj_event.asp) reference by W3Schools.

the `$(document)` selector, since we want to "listen to" mouse movement over the entire document:

```html
<!DOCTYPE html>
<html>
    <head>
        <script src="js/jquery.js"></script>
    </head>
    <body>
        <p id="position"></p>
        <script>
            $(document).on("mousemove", function(e) {
                $("#position").text(e.pageX + " " + e.pageY);
            });
        </script>
    </body>
</html>
```

The result is shown in Figure 4.8. The current mouse coordinates are displayed in the top-left corner of the screen. Every time the mouse is moved, the coordinates are updated.

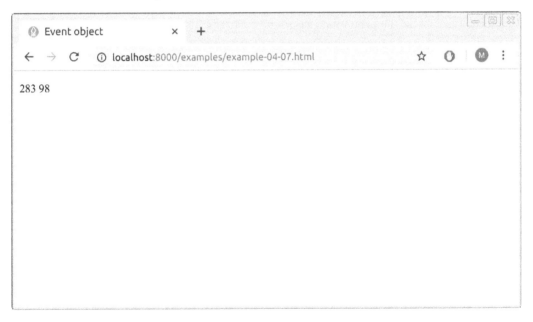

FIGURE 4.8: Screenshot of `example-04-07.html`

Different types of events are associated with different event object properties. Moreover, *custom* event types and event properties can be defined in JavaScript libraries. For example, later on in the book we will use the **map click** event to detect the clicked coordinates on an interactive map, then to trigger specific actions regarding that location. This is made possible by the fact that the `"click"` event on a map contains a custom property named `.latlng`, which can be accessed by the event listener function. We will see how this technique can be used for several purposes, such as:

- Displaying the clicked coordinates (Section 6.9)

- Adding a marker in the clicked location on the map (Section 11.2)
- Making a database query based on proximity to the clicked location (Section 11.4)

In Chapter 13, we will see another group of custom events referring to *drawing* and *editing* shapes on the map, such as `"draw:created"`, which is fired when a new shape is drawn on the map.

4.12   Iteration over arrays and objects

As mentioned earlier, the jQuery library has several functions for tasks other than selecting elements and operating on selection. One such function is `$.each`. The `$.each` function is a general function for **iterating** over JavaScript arrays and objects. This function can be used as a cleaner and shorter alternative to `for` loops (Section 3.10.3.2), in cases when we need to do something on each element of an array or an object[17]. The `$.each` function accepts two arguments:

- The **array** or **object**
- A **function** that will be applied on each element

The second argument, i.e., the function passed to `$.each`, also takes two arguments:

- `key`—The **index** (for an array) or the property **name** (for an object)
- `value`—The **contents** of the element (for an array) or the property **value** (for an object)

Like with e for "event" (Section 4.11), the parameter names `key` and `value` are chosen by convention. We could choose any pair of different names. The important point is that the first parameter refers to the index or property name, and the second parameter refers to the contents or the property value. For example, check out the following code:

```
var a = [52, 97, 104, 20];
$.each(a, function(key, value) {
    console.log("Index #" + key + ": " + value);
});
```

In the first expression, an array named a is defined. In the second expression, we are applying an (anonymous) function on each item of a. The function takes two arguments, `key` and `value`, and constructs a text string, which is printed in the console for each item. Note that the `$.each` method returns the array itself, which is why it is also being printed in the console.

- Open a web page where the jQuery library is included, such as `example-04-07.html`.
- Run the above code in the console and examine the printed output.
- Try rewriting the above code to use a `for` loop (Section 3.10.3.2), instead of `$.each`. (Hint: start with `for(var i in a)`.)

---

[17]The `.forEach` method is another plain-JavaScript alternative to `$.each`, though `.forEach` is less general—it only works on arrays, not on objects.

4.13 Modifying page based on data

One of the most important use cases of JavaScript is dynamically generating page contents based on *data*. The data we wish to display on the page can come from various sources, such as an object in the JavaScript environment, a file, or a database. JavaScript code can be used to process the data into HTML elements, which can then be added to the DOM (Section 4.7) and consequently displayed in the browser. Modifying page contents based on data is also a use case where iteration (Section 4.12) turns out to be very useful.

Our next example is a simple and small web page, containing just two paragraphs and one unordered list (Figure 4.9). We already know how to write the necessary HTML code for such a web page (Chapter 1). The novel part in `example-04-08.html` is that we are going to generate some of the HTML code based on data and using JavaScript, rather than have the entire HTML contents predefined.

Let's assume we need to dynamically create a list of items on our page, and we have an array with the contents that should go into each list item. We will use the jQuery `$.each` function for iterating over the array and adding its contents into an `<ul>` element on the page. For example, suppose we have the following array named `data`, including the Latin names of eight Oncocyclus *Iris* species[18] found in Israel:

```
var data = [
    "Iris atrofusca",
    "Iris atropurpurea",
    "Iris bismarckiana",
    "Iris haynei",
    "Iris hermona",
    "Iris lortetii",
    "Iris mariae",
    "Iris petrana"
];
```

In the HTML code, we may initially have an empty `<ul>` *placeholder*:

```
<ul id="species"></ul>
```

which we would like to fill with the `<li>` HTML elements (Section 1.6.7.1) based on the above array of species names[19], as follows:

```
<ul id="species">
    <li><i>Iris atrofusca</i></li>
    <li><i>Iris atropurpurea</i></li>
    <li><i>Iris bismarckiana</i></li>
    <li><i>Iris haynei</i></li>
    <li><i>Iris hermona</i></li>
```

---

[18]https://en.wikipedia.org/wiki/Iris_subg._Iris#Oncocyclus
[19]Note that we are also using the `<i>` element (Section 1.6.5.3) to display species in *italics*.

```
    <li><i>Iris lortetii</i></li>
    <li><i>Iris mariae</i></li>
    <li><i>Iris petrana</i></li>
</ul>
```

Why not embed the above HTML code directly into the HTML document, instead of constructing it with JavaScript? Two reasons why the former may not always be a good idea:

- The contents can be much longer, e.g., tens or hundreds of elements, which means it is less convenient to type the HTML by hand. We could use tools such as **Pandoc**[20] to programmatically convert text plus markup to HTML, thus reducing the need to manually type all of the HTML tags, but then it is probably more flexible to do the same with JavaScript anyway.
- We may want to build page content based on *real-time* data, loaded each time the user accesses the website, and/or make it responsive to user input. For example, a list of headlines in a news website can be based on a real-time news stories database and/or customized according to user preferences.

To build the above HTML code *programmatically*, based on the `data` array, we can use the `$.each` method we have just learned (Section 4.12). First, recall that `$.each` accepts a function that can do something with the `key` and `value` of each array element, such as collecting them in a two-element array and printing them into the console:

```
$.each(data, function(key, value) {
    console.log([key, value]);
});
```

Check out the printed output (below). It includes eight arrays. In each array, the first element is the `key` (the index: 0, 1, ...) and the second element is the `value` (the species name string).

```
[0, "Iris atrofusca"]
[1, "Iris atropurpurea"]
[2, "Iris bismarckiana"]
[3, "Iris haynei"]
[4, "Iris hermona"]
[5, "Iris lortetii"]
[6, "Iris mariae"]
[7, "Iris petrana"]
```

- Assign the `directory` object from Section 3.10.3.2 to a variable named `data` and run the above `$.each` iteration.
- How did the printed `key` and `value` change?
- Repeat the exercise with `directory.musicians`. What happened now?

---

[20]https://pandoc.org/

Now that we know how to iterate over our array **data**, we just need to modify the function that is being applied. Instead of just printing the array contents in the console, we want the function to create new **<li>** elements in the DOM. Additionally, we need to create the empty **<ul>** placeholder element in our document (using HTML) that the **<li>** elements will be appended to, as shown above.

To create an empty **<ul>** element, we can add the following HTML code inside the **<body>**:

```
<ul id="species"></ul>
```

Note that we are using an ID to identify the particular **<ul>** element, in this case **id="species"**. This is very important since we need to identify the particular element which our JavaScript code will operate on!

Next, to add a new **<li>** element *inside* the **<ul>** element, we can use the **.append** method, which inserts specified content at the end of the selected element(s) (Section 4.7.4). For example, the following expression will add one **<li>** element at the end of the list:

```
$("#species").append("<li><i>A list item added with jQuery<i></li>");
```

- Open **example-04-08.html** in the browser.
- Run the above expression in the console several times, to see more and more items being added to the list.

Instead of adding just one **<li>** item, what we actually need is to *iterate* over the **data** array, each time adding a **<li>** element as the last child of the **<ul>**. Replacing the constant **<li>** with a dynamic one, based on current value, and encompassing it inside a **$.each** iteration, the final code for dynamic construction of the list looks like this:

```
$.each(data, function(key, value) {
    $("#species").append("<li><i>" + value + "</i></li>");
});
```

The above means we are iterating over the **data** array, each time adding a new **<li>** at the bottom of our list, with the contents being the current **value** (displayed in italics using **<i>**). Here is the complete code of the small web page **example-04-08.html**, implementing the dynamic creation of an unordered list using JavaScript:

```
<!DOCTYPE html>
<html>
    <head>
        <title>Populating list</title>
        <script src="js/jquery.js"></script>
    </head>
```

```html
<body>
    <p>List of rare <i>Iris</i> species in Israel:</p>
    <ul id="species"></ul>
    <p>It was created dynamically using jQuery.</p>
    <script>
        var data = [
            "Iris atrofusca",
            "Iris atropurpurea",
            "Iris bismarckiana",
            "Iris haynei",
            "Iris hermona",
            "Iris lortetii",
            "Iris mariae",
            "Iris petrana"
        ];
        $.each(data, function(key, value) {
            $("#species").append("<li><i>" + value + "</i></li>");
        });
    </script>
</body>
</html>
```

The result is shown in Figure 4.9.

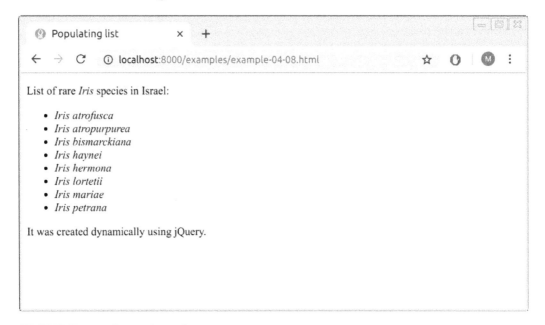

FIGURE 4.9: Screenshot of `example-04-08.html`

Again, in this small example the advantage of generating the list with JavaScript, rather than writing the HTML code by hand, may not be evident. However, this technique is very powerful in more realistic settings, such as when our data is much larger and/or needs to be constantly updated. For example, later on we will use this technique to dynamically generate a dropdown list with dozens of plant species names according to real-time data coming from a database (Section 10.4.4).

4.14 Working with user input

Our last example in this chapter concerns dynamic responses to user input (Section 1.6.12) on the page. In **example-04-09.html**, we will build a simple calculator app where clicking on a button multiplies two numbers the user entered, and prints the result on screen (Figure 4.10). In fact, we already covered everything we need to know to write the code for such an application. The only thing different from the previous examples, is that we are going to use the .**val** method (Section 4.7.7) to query the currently entered user input.

As usual, we start with HTML code. Importantly, our code contains three **<input>** elements:

- The first number (**type="number"**)
- The second number (**type="number"**)
- The "Multiply!" button (**type="button"**)

Here is the HTML code for just those three **<input>** elements:

```
<input type="number" id="num1" min="0" max="100" value="5">
<input type="number" id="num2" min="0" max="100" value="5">
<input type="button" id="multiply" value="Multiply!">
```

Note that all input elements are associated with IDs: **num1**, **num2**, and **multiply**. We need the IDs for referring to the particular elements in our JavaScript code. Below the **<input>** elements, we have an empty **<p>** element with **id="result"**. This paragraph will hold the multiplication result. Initially, the paragraph is empty, but it will be filled with content using JavaScript code:

```
<p id="result"></p>
```

The only scenario where the contents of the page change is when the user clicks on the "Multiply!" button. Therefore, our **<script>** is in fact composed of just one event listener, responding to **"click"** events on the "Multiply!" button:

```
$("#multiply").on("click", function() {...});
```

The function being passed to the event listener modifies the text contents of the **<p>** element. The inserted text gives the multiplication result, using the .**val** method to extract both input numbers, **$("#num1").val()** and **$("#num2").val()**:

```
$("#multiply").on("click", function() {
    $("#result")
        .text("The result is: " +  $("#num1").val() * $("#num2").val());
});
```

It is important to note that all input values are returned as text *strings*, even when we are talking about numeric `<input>` elements. You can see this by opening `example-04-09.html` and typing `$("#num1").val()` in the console. However, when using the multiplication operator `*` strings are automatically converted to numbers, which is why multiplication still works[21]. Here is the complete code of our arithmetic web application, given in `example-04-09.html`:

```html
<!DOCTYPE html>
<html>
    <head>
        <title>Working with user input</title>
        <script src="js/jquery.js"></script>
    </head>
    <body>
        <p>First number
      <input type="number" id="num1" min="0" max="100" value="5">
        </p>
        <p>Second number
      <input type="number" id="num2" min="0" max="100" value="5">
        </p>
        <input type="button" id="multiply" value="Multiply!">
        <p id="result"></p>
        <script>
            $("#multiply").on("click", function() {
                $("#result").text(
                    "The result is: " + $("#num1").val() * $("#num2").val()
                );
            });
        </script>
    </body>
</html>
```

The result is shown in Figure 4.10.

4.15   Exercise

- Modify `example-04-09.html` by adding the following functionality:
    - Make calculator history append at the top, to keep track of the previous calculations. For example, if the user multiplied 5 by 5, then 5*5=25 will be added on top of all other history items. If the user made another calculation, it will be added above the previous one, and so on (Figure 4.11).
    - Add a second button for making *division* of the two entered numbers.
    - Add a third button for *clearing* all previous history from screen. (Hint: you can use `.html("")` to clear the contents of an HTML element.)

---

[21]Try typing `"5"*"5"` in the console to see this behavior in action.

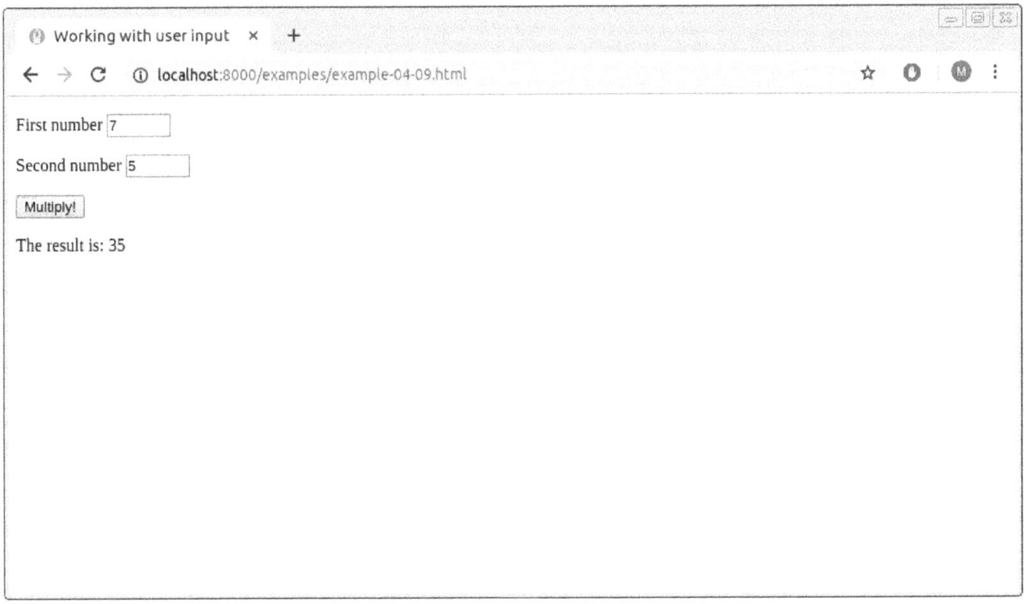

FIGURE 4.10: Screenshot of `example-04-09.html`

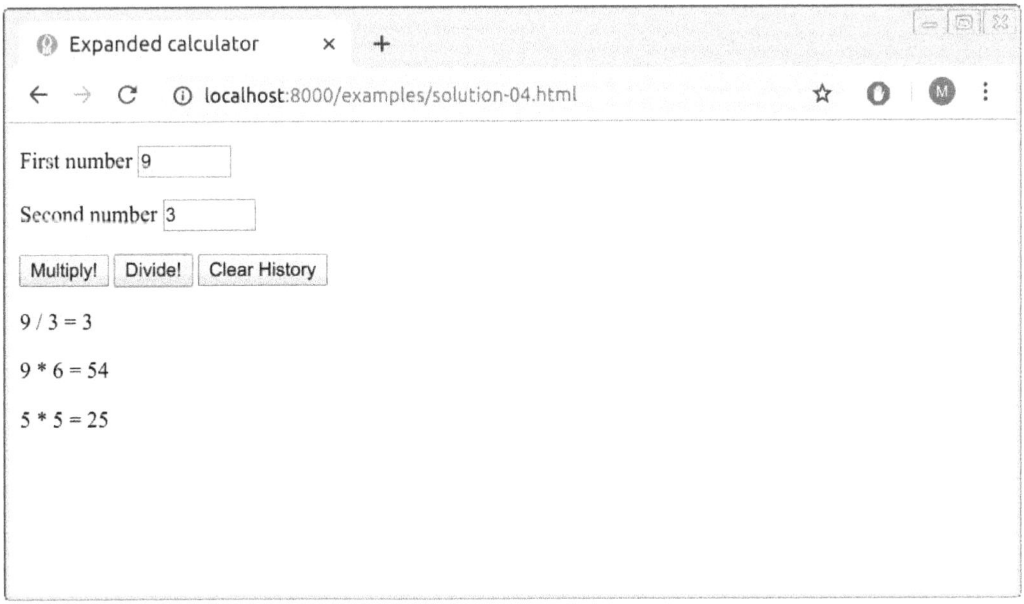

FIGURE 4.11: Screenshot of `solution-04.html`

5

Web Servers

5.1 Introduction

So far we have primarily dealt with the code—HTML (Chapter 1), CSS (Chapter 2), and JavaScript (Chapter 3–4)—required for building a website. We have not yet considered, however, the infrastructure required for *hosting* a website on a server, and the nature of *communication* between that server and the client, the web browser.

As mentioned in Chapter 1, at the most basic level a web page is an HTML document which is located at a node on the internet. This node is called a **server**[1], as it serves the file to the world wide web, allowing your computer, the **client**[2], to access it. When you open a web browser, such as Chrome, and enter a URL, such as `https://www.google.com/`, into the address bar, the web browser navigates to the node you have specified and requests this document, which it reads, interprets, and displays on our screen. The browser also applies CSS styling rules (Chapter 2) and runs JavaScript (Chapter 4), in case those are linked to the document.

This means that to host a website you need to take care of two things:

- You need to have the right kinds of documents and code files.
- You need to have a location on the internet (hardware) where you can place these documents and code files, as well as an appropriate environment (software) to *serve* them.

We already discussed what kind of documents and code files we can use to build a website. Specifically, we learned about:

- **HTML**, in Chapter 1
- **CSS**, in Chapter 2
- **JavaScript**, in Chapters 3–4

We have also seen how our customized HTML documents, along with any associated files, such as CSS and JavaScript, can be created using a plain text editor, such as Notepad++, then opened and viewed in the browser. The natural question that arises is what do we need to do to make the next step and turn our page into a "real" website, one that can be accessed and viewed by other people rather than just us.

In this chapter we focus on exactly that, the second part of the picture: having a location where our documents are placed, and software to publish or serve them over the network, so that they are accessible to other people.

---

[1] https://en.wikipedia.org/wiki/Web_server
[2] https://en.wikipedia.org/wiki/Client_(computing)

5.2   Web servers

The term web server can refer to hardware or software, or both of them working together, for serving content over the internet. On the *hardware* side, a web server is a computer that stores a website's component files, such as:

- HTML documents
- CSS stylesheets
- JavaScript files
- Other types of files, such as images

The server delivers these files to the client's device. It is connected to the internet and can be accessed through a URL such as `https://www.google.com/`.

On the *software* side, a web server includes several parts that control how web users access the hosted files. The minimal required software component is an **HTTP server**. An HTTP server is a software component that understands URLs (web addresses) and **Hypertext Transfer Protocol (HTTP)**[3]—the protocol used by the browser to communicate with the server (Section 5.3). A server with just the HTTP server component is referred to as a *static* server (Section 5.4.2), as opposed to a *dynamic* server (Section 5.4.3) which has several additional components.

At the most basic level, whenever a browser needs a file hosted on a web server, the browser requests the file via the HTTP protocol (Section 5.3). When the request reaches the correct web server (hardware), the HTTP server (software) sends the requested document back, also through HTTP (Figure 5.1).

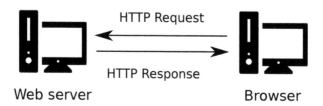

FIGURE 5.1: Client-server communication through HTTP

Don't worry if this is not clear yet—we elaborate on HTTP servers, the HTTP protocol, URLs, static and dynamic servers, in the following Sections 5.3–5.4.

5.3   Communicating through HTTP

5.3.1   Web protocols and HTTP

HTTP is a protocol specifying the way that communication between the client and the server takes place. As its name—Hypertext Transfer Protocol—implies, HTTP is mainly used to transfer hypertext (i.e., HTML documents) between two computers.

---

[3]`https://en.wikipedia.org/wiki/Hypertext_Transfer_Protocol`

HTTP is not the only protocol in use for communication on the web. For example, **FTP**[4] and **WebSocket**[5] are examples of other web communication protocols. However, HTTP is the most basic and most commonly used protocol. Almost everything we do online, including everything we do in this book, is accomplished through HTTP communication. The secured version of HTTP, known as **HTTPS**[6], is becoming a very common alternative to HTTP and thus should be mentioned. However, HTTPS just adds a layer of security through encrypted communication and is not fundamentally different from HTTP.

In the context of the web, a protocol is a set of rules for communication between two computers. HTTP, specifically, is a textual and stateless protocol:

- **Textual** means that all commands are plain-text, therefore human-readable.
- **Stateless** means that neither the server nor the client remember previous communications. For example, an HTTP server, relying on HTTP alone, cannot remember if you are "logged-in" with a password, or in what step you are in a purchase transaction.

Furthermore, only clients can make HTTP requests, and then only to servers. Servers can only respond to a client's HTTP request. When requesting a file via HTTP, clients must provide the file's URL. The web server must answer every HTTP request, at least with an error message. For example, in case the requested file is not found the server may return the "404 Not Found"[7] error message. The 404 error is so common that many servers are configured to send a customized 404 error page.

- Try navigating to `https://google.com/abcde.html`, or any other non-existing document on `https://google.com/`.
- What you see is Google's customized 404 error page.

5.3.2 HTTP methods

5.3.2.1 Overview

The HTTP protocol defines several *methods*, or verbs, to indicate the desired action on the requested resource on the server. The two most commonly used HTTP methods, for a request-response between client and server, are `GET` and `POST`.

- `GET`—Used to **request** data from the server (Section 5.3.2.2)
- `POST`—Used to **submit** data to be processed on the server (Section 5.3.2.3)

There are a few other methods[8], such as `PUT` and `DELETE`, which are used much more rarely and we will not go into.

---

[4]https://en.wikipedia.org/wiki/File_Transfer_Protocol
[5]https://en.wikipedia.org/wiki/WebSocket
[6]https://en.wikipedia.org/wiki/HTTPS
[7]https://en.wikipedia.org/wiki/HTTP_404
[8]https://en.wikipedia.org/wiki/Hypertext_Transfer_Protocol#Request_methods

5.3.2.2   The GET method

The GET method is used to request data. It is by far the most commonly used method in our usual interaction with the web. For example, typing a URL in the address bar of the browser in fact instructs the browser to send a GET request to the respective server. A static server (Section 5.4.2) is sufficient for processing GET requests in case the requested file is physically present on the server. The response is usually an HTML document, which is then displayed in the browser, but it can also be other types of content, such as GeoJSON (Section 3.11.2).

In addition to manual typing in the browser address bar, GET requests can also be sent programmatically, by running code. In this book, we will frequently send GET requests using JavaScript code. For example, in the following chapters we will learn about a method for loading GeoJSON content from local files (Section 7.8.2), or from remote locations on the web (Sections 7.8.3 and 9.7), using GET requests.

5.3.2.3   The POST method

The POST method is used when the client sends data to be processed on the server. It is more rarely used compared to GET, and somewhat more complicated. For example, there is no way to send a POST request by simply typing a URL in the browser address tab, unlike with GET. Instead, making a POST request to a web server can only be made through code, such as JavaScript code. Also, a dynamic server (Section 5.4.3) is required to process POST requests, where server-side scripts determine what to do with the received data. Plainly speaking, POST requests are preferred over GET requests when we need to send substantial amounts of data to the server.

In this book, we will encounter just one example of using POST requests, in Chapter 13. In that chapter, we will build a crowdsourcing web application where the user draws layers on a web map. These layers are subsequently sent for permanent storage in a database. Sending the drawn layer to the server (Section 13.6) will be accomplished using POST requests.

5.4   Static vs. dynamic servers

5.4.1   Overview

Web servers can be divided in two general categories:

- **Static** web servers
- **Dynamic** web servers

What we discussed so far, and what we use in this book, refers to static servers. Dynamic servers have some additional complexity, and we will only mention them, for general information, in Section 5.4.3.

5.4.2 Static servers

As noted previously, a **static server**[9] consists of a computer (hardware) with just an HTTP server (software). We call it "static" because the server sends its hosted files "as-is" to your browser, without any additional pre-processing. This means the static server can only respond to GET requests (Section 5.3.2.2) for pre-existing HTML documents, and send those documents to the browser. While loading the HTML document, the browser may send further GET requests for other pre-existing files linked in the HTML code, such as CSS, JavaScript, images, and so on.

For example, suppose we are vising a hypothetical website focused on travel locations, and we are navigating to a specific page on travelling to France, at http://www.travel.com/locations/france.html. In case the website is served using a static server, there is an actual france.html document on the server. All the server has to do is send you a copy of that file (Figure 5.2).

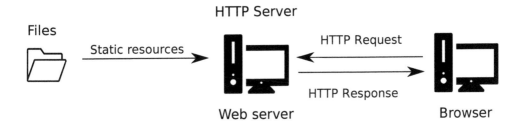

FIGURE 5.2: Static server architecture

As another example, the online version of this book (Section 0.7)—which you are, perhaps, reading at the moment—is hosted on a static server. This means that all of the HTML documents comprising the website are prepared in advance. Entering a URL for a specific page (such as web-servers-1.html, for *this* chapter) sends the appropriate file to your browser through HTTP.

5.4.3 Dynamic servers

A **dynamic server**[10] consists of an HTTP server plus extra software, most commonly an application server and a database. We call it "dynamic" because the server dynamically builds the HTML documents, or any other type of content, before sending them to your browser via the HTTP server. Typically, the dynamic server uses its application server, i.e., software running server-side scripts, HTML templates, and a database to assemble the HTML code. Once assembled, the dynamically assembled HTML content is sent via HTTP, just like static content.

With a dynamic server, when entering the above-mentioned hypothetical URL, http://www.travel.com/locations/france.html, into the address bar in our browser, the france.html document *doesn't exist yet*. The server waits for your request, and when the

---

[9]https://en.wikipedia.org/wiki/Static_web_page
[10]https://en.wikipedia.org/wiki/Dynamic_web_page

request comes in, it uses various "ingredients" (e.g., templates, a database, etc.) and a "recipe" to create that page on the spot, just for you (Figure 5.3).

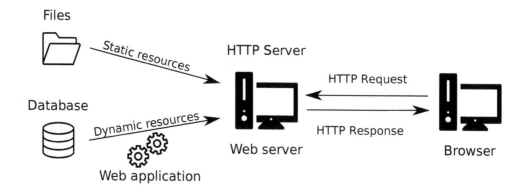

FIGURE 5.3: Dynamic server architecture

For example, websites like Wikipedia[11] are composed of many thousands of separate web pages, but they aren't real HTML documents, only a few HTML templates and a giant database[12]. This setup makes it easier and quicker to maintain and deliver the content[13].

5.4.4   Software

As we will see shortly (Section 5.6), running a static server is easy to do on our own—assuming we already have our web page documents prepared. An HTTP server is included in many software packages and libraries, and it does not require any special installation or configuration. There are numerous software solutions which can start an HTTP server in a few minutes; **Python**[14] (Section 5.6.2) and **R**[15] are just two examples. There are also several free cloud-based options to have a managed static server such as **GitHub Pages** (Section 5.6.3), which means you do not even need to have your own dedicated computer or invest in paid cloud-based services to run your static server. There are also professional HTTP server software packages, used for building both static and dynamic servers, which we will not use in this book. At the time of writing, two most commonly used ones are **Apache HTTP Server**[16] and **Nginx**[17].

Setting up and running a dynamic server is more complicated, requiring specialized installation, configuration, and maintenance. There are no instant solutions, such as the ones we will see shortly for static servers (Section 5.6), since it is up to us to define the way in which the server dynamically generates HTML content. The latter is done by setting up an application server and writing custom **server-side scripts**[18]. With a dynamic server, in

---

[11]https://en.wikipedia.org

[12]https://en.wikipedia.org/wiki/Wikipedia#Hardware_operations_and_support

[13]More information on the differences between static and dynamic servers can be found in the *Introduction to the server side* article (https://developer.mozilla.org/en-US/docs/Learn/Server-side/First_steps/Introduction) by Mozilla.

[14]https://docs.python.org/3/library/http.server.html

[15]https://cran.r-project.org/package=servr

[16]https://en.wikipedia.org/wiki/Apache_HTTP_Server

[17]https://en.wikipedia.org/wiki/Nginx

[18]https://en.wikipedia.org/wiki/Server-side_scripting

addition to the HTTP server software, you need to write server-side scripts which run on the server, as opposed to client-side JavaScript scripts that run on the client (Chapter 4). Server-side scripts are responsible for tasks such as generating customized HTML content, authentication, managing user sessions, etc. There are several programming languages (and frameworks) that are commonly used for writing server-side scripts, such as **PHP**[19], **Python (Django)**[20], **Ruby (on Rails)**[21], and **JavaScript (Node.js)**[22].

5.4.5 Practical considerations

There are advantages and disadvantages to both the static and the dynamic server approaches. *Static sites* (i.e., sites served with a static server) are simple, fast, and cheap, but they are harder to maintain (if they are complex) and impersonal. *Dynamic sites* provide more flexibility and are easier to modify, but also slower, more expensive, and technically more difficult to build and handle.

In this book, we will only build static sites hosted using a static web server. A static server cannot use a database or template to send personalized HTML content, just pre-compiled HTML documents. Nevertheless, as we will see throughout the book, a static server is not limited to showing fixed, non-interactive content. For example, the HTML content of the web page can be modified in response to user actions through client-side scripts (in JavaScript), without needing a server, using the methods we learned in Chapter 4. Later on, we will also see that static pages can dynamically "grab" information from other locations on the web, again using client-side JavaScript, including from existing dynamic servers and databases (Section 9.7). That way, we can integrate dynamic content even though we do not operate our own dynamic server. Still, there are things that can only be accomplished with a dynamic server.

The most notable example where a dynamic server is an obvious (and only) solution is **authentication**. For example, suppose we want to create a password-protected website. In our website, we can add a form with an input element (Section 1.6.12) where the user enters a password, and only if the password is valid—the content will be shown. This requires authentication—some way to evaluate the validity of the entered password. Now, suppose we have a database of valid passwords for the various authorized website users. Where can we place that database, and how can our page access it? If we place it directly on the client, e.g., as an array (Section 3.6.3.1) in our JavaScript code, the contents will be exposed to anyone looking at the source code of our page. Remember that whenever a JavaScript script is linked to our HTML document, the user can access and view the code. Even if we place the password database on a different location, such as a separate static server, we still need to hard-code the instructions for accessing that other location in our JavaScript code, so that the web page can access it. Again, anyone who reads those instructions can access the database the same way the browser does. The solution is to send the user-entered password for validation using a server-side script. If the password is valid, the server can return an "OK" message and/or any content that the specific user is allowed to see. That way, the password database is not accessible, since the server is not allowed to send it—only to accept an entered password and compare it to those in the database.

From now on, we concentrate on static servers and how to set them up.

---

[19]https://en.wikipedia.org/wiki/PHP
[20]https://en.wikipedia.org/wiki/Django_(web_framework)
[21]https://en.wikipedia.org/wiki/Ruby_on_Rails
[22]https://en.wikipedia.org/wiki/Node.js

5.5 URLs and file structure

5.5.1 URLs and `index.html`

As we will see in a moment (Section 5.6), a static server is associated with a directory on the computer, serving its contents over the web. Once the server is running, a client can request any HTML document (or other type of file) which is located inside that directory, or any of its sub-directories, by entering a **Uniform Resource Locator (URL)** in the browser address bar. To construct the right URL for accessing a given HTML document (or other resource), one needs to know two things:

- The **IP address**[23] of the host computer and the **port**[24] where the server is running, or, alternatively, the **domain name**[25] (see below)
- The **path** to the HTML document on the server

For example, the online version (Section 0.7) of Chapter 5, which you are reading right now, can be reached by entering the following URL into the browser address bar:

`http://159.89.13.241:8000/web-mapping/web-servers-1.html`

Let's go over the separate components this URL is composed of:

- `http://` means we are communicating using **HTTP**. This part is automatically completed by the browser, so it can be omitted.
- `159.89.13.241` is the **IP address** of the web server that hosts the website. The IP address is a unique identifier given to a computer connected to a network, making it possible to identify the computer in the network.
- `:8000` is the **port** number where the server is running (Section 5.6.2.2). When using the default port for a given communication protocol, which is 80 for HTTP and 443 for HTTPS, the port number can be omitted.
- `/web-mapping/web-servers-1.html` is the location of the **document**. With a static server, this means that within the directory we are serving there is a sub-directory named `web-mapping`, and inside it there is an HTML document named `web-servers-1.html`[26].

What happens if we remove the last part—the HTML file name `web-servers-1.html`—from the URL, thus navigating to the `/web-mapping/` *directory* instead?

`http://159.89.13.241:8000/web-mapping/`

Try it, and you should see the `index.html` page (the book's Preface), even though we did not specify any HTML file name. This happens because standard protocol dictates that a file named `index.html` will be provided by default when we navigate to a directory, rather than an HTML document, on the web server. The `index.html` file usually contains the first page users see when navigating to a website.

You may now be wondering how come we are usually navigating to a *textual* URL such as `https://www.google.com`, rather than a *numeric* IP address and port number, such as `http://216.58.206.14:80/`? The answer is something called a **Domain Name Server**

---

[23]`https://en.wikipedia.org/wiki/IP_address`
[24]`https://en.wikipedia.org/wiki/Port_(computer_networking)`
[25]`https://en.wikipedia.org/wiki/Domain_name`
[26]More information on URL structure can be found in the *What is a URL?* article (`https://developer.mozilla.org/en-US/docs/Learn/Common_questions/What_is_a_URL`) by Mozilla.

(DNS)[27]. When you enter a textual URL into your browser, the DNS uses its resources to resolve the domain name into the IP address for the appropriate web server. This saves us the trouble of remembering IP addresses, using more recognizable textual addresses instead.

5.5.2 File structure

The following diagram shows a hypothetical file structure of a static website directory:

```
|-- www
    |-- css
        |-- style.css
    |-- images
        |-- cat.jpg
    |-- js
        |-- main.js
    |-- dog.jpg
    |-- index.html
```

The various files that comprise the website are located either in the **root directory** or in **sub-directories** within the root directory. In the above example, `www` represents the root directory of the website. In this example, the root directory `www` contains a default `index.html` document. As mentioned above (Section 5.5.1), this means that when we browse to the directory address without specifying a file name, the `index.html` file is sent by default. The root directory `www` also contains sub-directories. Here, the sub-directories are used for storing additional files linked to the hypothetical HTML code of `index.html`:

- `css`—for CSS files (`.css`)
- `images`—for images
- `js`—for JavaScript files (`.js`)

The structure and names of the sub-directories are entirely up to us, web page developers. We can even place all of the files in the root directory, without any internal division to sub-directories. However, it is usually convenient to have the type of sub-directory structure as shown above, where files of different types are stored in separate sub-directories. That way, the various components that make up our website can be easier to track and maintain.

5.5.3 Relative paths

In the above file structure example, the `images` folder contains an image file named `cat.jpg`. In case we want this image to be displayed on the `index.html` page, the HTML code in `index.html` needs to include an `<img>` element (Section 1.6.9). The `src` attribute of that `<img>` element needs to refer to the `cat.jpg` file. Either one of the following versions will work:

```
<img src="/images/cat.jpg">
<img src="images/cat.jpg">
```

---

[27]https://developer.mozilla.org/en-US/docs/Learn/Common_questions/What_is_a_domain_name

Both versions of the `src` attribute value are known as *relative* file paths, because they are relative to a given location on the server:

- In the first case, the path is relative to the **root** directory of the server, specified by the initial / symbol.
- In the second case, the path is relative to the **current** directory where the HTML is loaded from, so the path just starts with a directory or file name in the same location (without the / symbol).

In this particular example the `index.html` file is in the root directory, thus the current directory is identical to the root directory. Therefore the `cat.jpg` file can be reached with either `/images/cat.jpg` or `images/cat.jpg`. Incidentally, we have another image file `dog.jpg` in the root directory (which, again, is the current directory for `index.html`). We can insert the `dog.jpg` image into the `index.html` document using either of the following `<img>` elements:

```
<img src="/dog.jpg">
<img src="dog.jpg">
```

This is another example of using paths that are relative to the *root* or *current* directory, respectively.

- In case the `index.html` file was moved (for some reason) from the root directory to the `images` directory, the `src` attribute in the `<img>` element for loading `cat.jpg` would have to be changed to `src="/images/cat.jpg"` or to `src="cat.jpg"`.
- Can you explain why?

5.5.4   CSS and JavaScript

5.5.4.1   Overview

As mentioned earlier (Sections 2.7.4 and 4.5.2), CSS and JavaScript code can be loaded from separate files, usually ending in `.css` and `.js`, respectively. This takes a little more effort than including CSS and JavaScript inside the HTML document, but saves work as our sites become more complex, because:

- Instead of repeating the same CSS and JavaScript code in different pages of the website, we can load the same file in all pages.
- When modifying our external CSS or JavaScript code, all web pages loading those files are immediately affected.

5.5.4.2   Linking CSS

Linking an external CSS file can be done using the `<link>` element within the `<head>` of the HTML page (Section 2.7.4). For example, in our hypothetical static server file structure

(Section 5.5.2), the `css` folder contains a `style.css` file. This CSS file can be linked to the `index.html` document be including the following `<link>` element:

```
<link rel="stylesheet" href="/css/style.css" type="text/css">
```

Note that, in this case, it makes sense to use a relative path which is relative to the root directory, since a website usually has a single set of CSS and JavaScript files. That way, exactly the same `<link>` element can be embedded in all HTML documents, regardless of where those HTML documents are placed.

5.5.4.3 Linking JavaScript

Linking a JavaScript code file can be done by adding a file path in the `src` attribute of a `<script>` element (Section 4.5.2). The `<script>` element can then be placed in the `<head>` or the `<body>` of the HTML document. For example, our hypothetical file structure has a folder named `js` with a JavaScript file named `main.js`. That script can be loaded in the `index.html` document by including the following `<script>` element:

```
<script src="/js/main.js"></script>
```

Again, a relative path, relative to the root directory, is being used.

5.6 Running a static server

5.6.1 Overview

So far we have discussed several background topics related to running a static server:

- Communication through HTTP (Section 5.3)
- Difference between static and dynamic servers (Section 5.4)
- Components of a URL and the file structure on the server (Section 5.5)

What is left to be done is actually *running* a server, to see how it all works in practice. In the next two sections we will experiment with running a static web server using two different methods:

- A **local server**, using your own computer and Python (Section 5.6.2)
- A **remote server**, using the GitHub Pages platform (Section 5.6.3)

5.6.2 Local with Python

5.6.2.1 Setup instructions

We begin with the local option for running a static server. The exercise will demonstrate the HTTP server built into Python, which only requires you have **Python** installed. If you are working in a computer classroom, there is a good chance that Python is already installed. In any case, you can check if Python is installed by opening the **Command Prompt** (open the **Start** menu, then type `cmd`) and typing `python`. If you see a message with the Python version number, then Python is installed and you have just entered its command line, marked by the `>>>` symbol. You can exit the Python command line by typing `exit()` and pressing **Enter**. In case you see an error message, such as the following one, then Python is not installed:

```
'python' is not recognized as an internal or external command,
operable program or batch file.
```

Python installation instructions are beyond the scope of the book, but there is a plenty of information on installing Python that you can find online.

5.6.2.2 Running the server

To run Python's HTTP Server, follow these steps:

- Open the **Start** menu and type `cmd` to enter the Command Prompt.
- Navigate to the directory (Section 5.5.2) that you want to serve (e.g., where your `index.html` file is), using `cd` (change directory) followed by directory name. For example, if your directory is in drive `D:\`, inside a directory named `Data` and then a sub-directory named `server`, you need to type `cd D:\Data\server` to navigate to that directory.
- Type the expression `python -m SimpleHTTPServer` (if you are using **Python 2**) or `python -m http.server` (if you are using **Python 3**). To check which version of Python you have, type `python -V` in the Command Prompt.

If all goes well, you should see a message such as the following (Figure 5.4), meaning that the server is running:

```
Serving HTTP on 0.0.0.0 port 8000 ...
```

As evident from the above message, the default port (Section 5.5.1) where the Python server is running is 8000. In case you want to use a different port, such as 8080, you can specify the port number as an additional parameter:

```
python -m SimpleHTTPServer 8080   # Python 2
python -m http.server 8080        # Python 3
```

To stop the server, press **Ctrl+C**.

5.6.2.3 Testing served page

Once the server is running, you can access the served web page(s) by navigating to the following address in a web browser, assuming there is an `index.html` file in the root of the served directory:

```
http://localhost:8000/
```

In case you want to load an HTML document other than `index.html`, or if the document is located in one of the sub-directories of the server, you can specify the path to the HTML file you wish to load. For example:

```
http://localhost:8000/locations/france.html
```

If you initiated the server on a different port, replace the 8000 part with the port number you chose:

`http://localhost:8080/locations/france.html`

The word `localhost` means you are accessing *this computer*[28]. In other words, the server and the client are the same computer. This kind of setting may seem strange, but in fact it is extremely useful for development and testing of websites. Having the client and server on the same computer means that we can simulate and test the way that a client accesses the website, without needing to deploy the website on a *remote* server which requires some more work (Section 5.6.3).

Python's HTTP server prints incoming requests in the console, which is useful for monitoring how the server works. For example, the screenshot in Figure 5.4 shows several logged messages printed while using the server. In this particular printout, we can see that the server successfully processed four `GET` requests, all of which took place on February 22, in 2018, at around 15:08 (Figure 5.4).

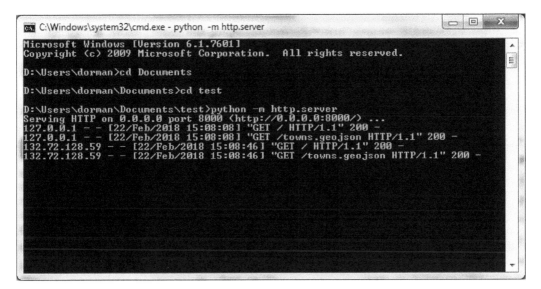

FIGURE 5.4: Running Python's simple HTTP server

5.6.2.4 Interactive map example

The following two exercises will demonstrate the concept of running a static server, using Python's HTTP server. In the first exercise, we will run the HTTP server to serve a web page with an interactive map of towns in Israel `example-08-07.html` (Figure 8.9). The purpose of this exercise is to become familiar with organization of multiple files in a static server directory. Don't worry if you don't understand the code that is *in* the files—it will be explained in detail later on (Section 8.8.2).

---

[28]https://en.wikipedia.org/wiki/Localhost

- Let's try using a static HTTP server to serve a web page consisting of multiple files over the network.
- Locate and download the file named `example-08-07.html` from the online version of the book (Section 0.7). Going over the source code, you will notice that the file is linked to one CSS file `css/leaflet.css` and two JavaScript files—`js/leaflet.js` and `js/jquery.js`. Additionally, the JavaScript code in the `<script>` of `example-08-07.html` loads a GeoJSON file `data/towns.geojson` using a method called Ajax (Section 7.7).
- Rename the `example-08-07.html` file to `index.html` and place it in an empty directory.
- Create sub-directories named `css`, `js` and `data` and place the appropriate files in each directory—`leaflet.css` in the `css` directory, `leaflet.js` and `jquery.js` in the `js` directory, and `towns.geojson` in the `data` directory.
- Start a local server from within the directory where the `index.html` file is placed, and open the page in the browser by navigating to `http://localhost:8000/`.
- You should see a map of town borders, with highlighted names on mouse hover (Figure 8.9).
- Now try opening the `index.html` file by double-clicking on it; the towns layer may now be absent, because loading GeoJSON content from a file using the Ajax (Section 7.7) method is blocked by some browsers (such as Chrome), unless running a server (Section 7.8.2). This demonstrates the necessity of going through the trouble of running a local server, for correctly emulating the way that web content is being served during website development.

5.6.2.5   Access from a different computer

In the second exercise, we will try to navigate to a web page served from a *different* computer. Note that this exercise will *not* work under certain network settings, due to different complications that require some more effort to overcome. For example:

- If you are connected to the internet through a **private network**, e.g., behind a router at your home, then the IP address of your computer (shown with `ipconfig`; see below) refers to an internal address of the private network, so other computers will not be able to reach your page simply by typing that IP address in the browser.
- If there is a **firewall** preventing other computers from reaching yours, then they will not be able to navigate to the page that you are serving.

The above considerations are handled by **network administrators**[29] and are beyond the scope of this book.

- While the page from the previous exercise is up and running, let's try to access it from a different computer over the network. This exercise should be done in pairs.
- Start up the static server with the interactive map from the previous example.
- Identify the IP address of your computer. To do that, click on the **Start** button, type `cmd` in the text box to open a *second* command line prompt (the first one should still

---

[29]https://en.wikipedia.org/wiki/Network_administrator

run your server so you cannot type additional commands there!), then type `ipconfig` in the command line. Locate your IP address in the printed output (Figure 5.5). The address should be listed next to the line where it says `IPv4 Address`. For example, in the output shown in Figure 5.5 the IP address is `132.72.129.98`.

- Tell the person next to you what your IP address is, and which port your server is running on. For example, if your IP address was `132.72.129.98` and the port number is 8000 then the address you should pass to the person next to you is `http://132.72.129.98:8000/`.
- The other person should type the address in his/her browser. If all worked well, your website will be displayed. Check your server's log—you should see the `GET` request(s) and the IP of the other computer that connected to your website!

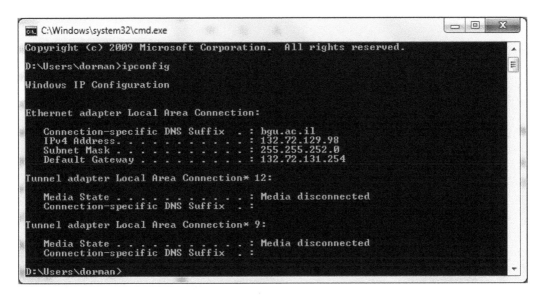

FIGURE 5.5: Determining the IP address using `ipconfig`

5.6.3 Remote with GitHub Pages

5.6.3.1 Overview

Python's HTTP server is simple enough to start working with, but there are other difficulties if you intend to use it for **production**, i.e., in real-life scenarios, where stability is essential. For instance, you need to take care of the above-mentioned network administration issues (Section 5.6.2.5), such as making sure your server has an IP address that can be reached from other computers, i.e., a public IP address, and that the server is not behind a firewall. In addition, you need to make sure the IP address of the computer always stays the same (a static IP address), take care of the hardware of your server, such as making sure the computer is always running and connected to the internet, make sure that the server is restarted in case the computer restarts, and so on. Using a remote hosting service, we basically let other people handle all of that. In other words, we don't need to worry for any of the hardware, software, and network connection issues—just the contents of our website.

There are numerous hosting services for static web pages. For example, both **Google**[30] and **Amazon**[31], as well as many other smaller companies, offer *paid* static hosting services. In this section, we will use the **GitHub** platform for hosting our static web page hosting, which is *free*. Although GitHub is mainly a platform for online storage of **Git** repositories and collaborative code development, one of its "side" functions is that of a static server. The static server functionality of GitHub is known as **GitHub Pages**. Using GitHub Pages as a remote static server has several advantages for our purposes:

- It is **simple**.
- It is **free**.
- It is part of **GitHub**, a popular platform for collaborative code development, which is useful to become familiar with.

Another good alternative is **surge.sh**[32]. It is also free, and can be quicker to set up compared to GitHub Pages, but requires using the command line.

5.6.3.2   Git and GitHub

We will not go into details on the functionality of GitHub, other than the GitHub Pages utility, which we use as a static server, but here is some background on what it is. When working with code, it becomes important to keep track of different versions of your projects. This allows you to undo changes made weeks or months ago. Versioning becomes even more important when collaborating with others, since in that case you may need to split your project into several "branches", or "merge" the changes contributed by several collaborators back together. To do all of those things, people use **version-control systems**. One of the most popular revision control systems around today is **Git**[33].

If Git is a version-control system, then what is **GitHub**[34]? Git projects are also called repositories. GitHub is a web-based Git repository hosting service. Basically, GitHub is an online service where you can store your Git repositories, either publicly or privately. The platform also contains facilities for interacting with other people, such as raising and discussing issues or subscribing to updates on repositories and developers you are interested in, creating a community of online code-collaboration. For anyone who wants to take part of open-source software development, using Git and GitHub is probably the most important skill after knowing how to write the code itself.

Importantly to our cause, for any public GitHub repository, the user can trigger the **GitHub Pages**[35] utility to serve the contents of the repository. As a result, the contents of the repository will be automatically hosted at the following address[36]:

```
https://GITHUB_USER_NAME.github.io/REPOSITORY_NAME/
```

where:

- `GITHUB_USER_NAME` is the **user name**
- `REPOSITORY_NAME` is the **repository name**

---

[30]https://cloud.google.com/storage/docs/hosting-static-website
[31]http://docs.aws.amazon.com/AmazonS3/latest/dev/WebsiteHosting.html
[32]https://surge.sh/
[33]https://git-scm.com/
[34]https://github.com/
[35]https://pages.github.com/
[36]Starting from 2016, GitHub Pages enforces HTTPS, which is why the address starts with `https://`. Typing `http://` is allowed but automatically redirects to `https://`.

Everything we learned about static servers applies in remote hosting too. The only difference is that the served directory is stored on another, remote server, rather than your own computer. For example, in order for a web page to be loaded when one enters a repository URL as shown above, you need to have an `index.html` file in the root directory of your GitHub repository (Section 5.5.1).

5.6.3.3 Setup instructions

What follows are step-by-step instructions for running a remote static server on GitHub Pages. To host our website on GitHub pages, go through the following steps:

- Create a GitHub account on `https://github.com/` (in case you don't have one already), then sign-in to your account. Your username will be included in all of the URLs for GitHub pages you create, as in `GITHUB_USER_NAME` in the URL shown above. In the screenshots in Figures 5.6–5.12, the GitHub username is `michaeldorman`.
- Once you are logged-in on `https://github.com/`, click the + symbol on the top-right corner and select **New repository** (Figure 5.6).
- Choose a name for your repository. This is the `REPOSITORY_NAME` part that users type when navigating to your site, as shown in the above URL. In the screenshots (Figures 5.6–5.12) the chosen repository name is `test`.
- Make sure the *Initialize this repository with a README* box is checked. This will create an (empty) `README.md` file in your repository, thus exposing the **Upload files** screen which we will use to upload files into our repository.
- Click the **Create repository** button (Figure 5.7).
- The newly created repository should be empty, except for one file named `README.md` (Figure 5.8).
- Click on the **Settings** tab to reach the repository settings.
- On the settings page, scroll down to the **GitHub Pages** section.

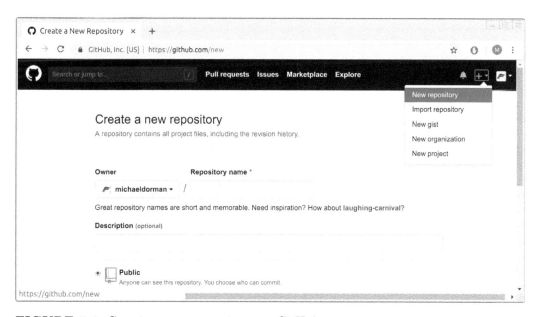

FIGURE 5.6: Creating a new repository on GitHub

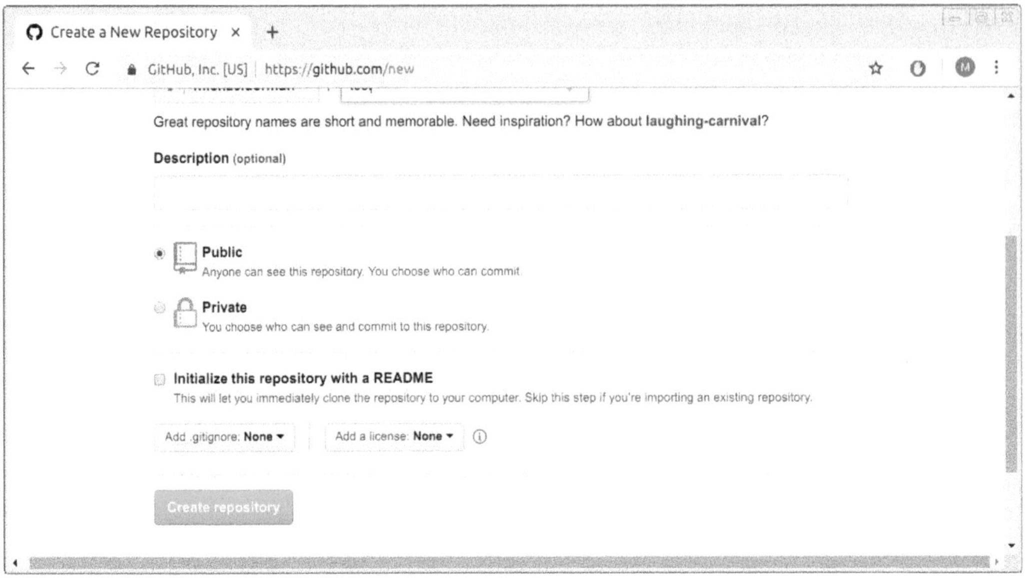

FIGURE 5.7: The *Create repository* button

FIGURE 5.8: Newly created repository, with `README.md` file

- In the **Source** panel, instead of **None** select **master branch** and click the **Save** button (Figure 5.9).
- Go back to the repository page and click the **Upload files** button. This will take you to the file upload screen (Figure 5.10).
- Drag and drop all files and folders that comprise your website into the box. This should usually include at least an HTML document named `index.html`.

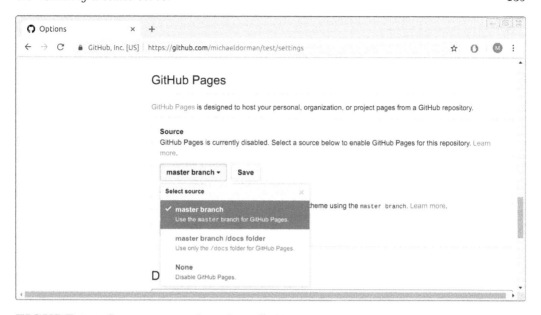

FIGURE 5.9: Setting `master` branch as GitHub Pages source

FIGURE 5.10: File upload screen

- Wait for the files to be transferred. Once all files are uploaded, click the **Commit changes** button (Figure 5.11).
- That's it! Your website should be live at `http://GITHUB_USER_NAME.github.io/REPOSITORY_NAME/`. Replace `GITHUB_USER_NAME` and `REPOSITORY_NAME` with your own GitHub user name and repository name, respectively[37].

---

[37]Note that you may have to wait a few moments before the site is being set up.

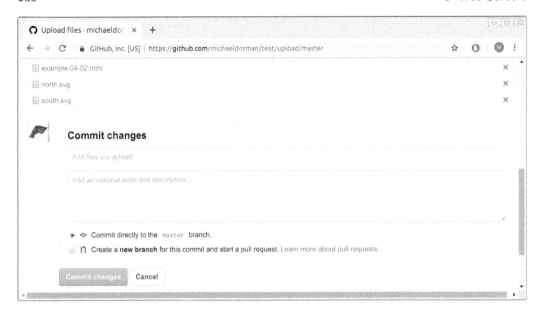

FIGURE 5.11: The *Commit changes* button

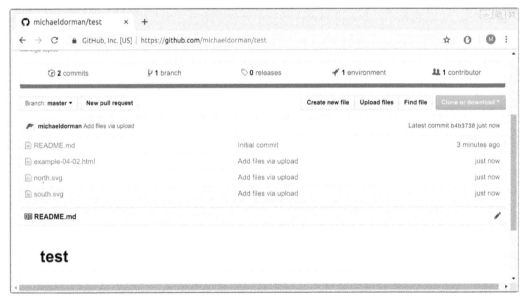

FIGURE 5.12: The repository with uploaded files

Part II

Web Mapping with Leaflet

6

Leaflet

6.1 Introduction

Now that we have covered the basics of web technologies, we are moving on to the main topic of this book: web mapping. This chapter and the next two (Chapters 6–8) introduce **Leaflet**, a JavaScript library used to create interactive web maps. Using Leaflet, you can create a simple map using just two or three JavaScript expressions, or you can build a complex map using hundreds of lines of code.

In this chapter, we will learn how to initialize a Leaflet web map on our web page, and how to add several types of layers on the map: tile layers (Section 6.5.7) and simple shapes such as point markers (Section 6.6.2), lines (Section 6.6.3), and polygons (Section 6.6.4). We will also learn to add interactive popups for our layers (Section 6.7) and a panel with a textual description of our map (Section 6.8). Finally, we will introduce map events—browser events associated with the web map (Section 6.9).

In the next two Chapters 7–8, we will learn some more advanced Leaflet functionality. In Chapter 7, we will learn to add complex shapes coming from external GeoJSON files. Then, in Chapter 8, we will learn how to define symbology and interactive behavior in our web map.

6.2 What is a web map?

We already introduced the concept of web mapping in Section 0.1. We mentioned that the term "web map" usually implies a map that is not simply *on* the web, but rather one that is *powered* by the web. It is usually interactive, and not always self-contained, meaning that it "pulls" content from other locations, such as tile layer servers (Section 6.5.8) or database APIs (Section 9.7)[1].

Similarly to spatial information displayed in GIS software[2], web maps are usually composed of one or more **layers**. Web map layers can be generally classified into two categories:

- **Background** layers, or basemaps, comprising collections of gridded images or vector tiles, which are usually general-purpose and not prepared specifically for our map

---

[1] A good introduction to web mapping can be found in the *Web Maps 101* (http://maptimeboston.github.io/web-maps-101/#0) presentation by *Maptime Boston* (https://maptimeboston.github.io/resources).

[2] https://en.wikipedia.org/wiki/Geographic_information_system

- **Foreground** layers, or overlays, which are usually vector layers (points, lines, and polygons), commonly prepared and/or fine-tuned for the specific map web where they are shown

Background layers are usually static and non-interactive. Conversely, foreground layers are usually dynamic and associated with user interaction, such as the ability to query layer attributes by clicking on a feature (Section 6.7).

6.3   What is Leaflet?

Leaflet[3] (Crickard III, 2014) is an open-source JavaScript library for building interactive web maps. Leaflet was initially released in 2011 (Table 6.1). It is lightweight, relatively simple, and flexible. For these reasons, Leaflet is probably the most popular *open-source* web-mapping library at the moment. As the Leaflet home page puts it, the guiding principle behind this library is simplicity:

"Leaflet doesn't try to do everything for everyone. Instead it focuses on making the basic things work perfectly."

Advanced functionality is still available through Leaflet plugins[4]. Towards the end of the book, we will learn about two Leaflet plugins: **Leaflet.heat** (Section 12.6) and **Leaflet.draw** (Section 13.3).

6.4   Alternatives to Leaflet

In this book, we will exclusively use Leaflet for building web maps. However, it is important to be aware of the landscape of alternative web-mapping libraries, their advantages and disadvantages. Table 6.1 lists Leaflet along with other popular JavaScript libraries for web mapping[5].

TABLE 6.1: Popular web-mapping libraries

Library	Released	Type	URL
Google Maps	2005	Commercial	`https://developers.google.com/maps/`
OpenLayers	2006	Open-source	`https://openlayers.org/`
ArcGIS API for JS	2008	Commercial	`https://developers.arcgis.com/javascript/`
Leaflet	2011	Open-source	`https://leafletjs.com/`
D3	2011	Open-source	`https://d3js.org/`
Mapbox GL JS	2015	Commercial	`https://www.mapbox.com/mapbox-gl-js/api/`

---

[3]`https://leafletjs.com/`
[4]`https://leafletjs.com/plugins.html`
[5]Check out the *What is a web mapping API?* (`https://www.e-education.psu.edu/geog585/node/763`) article for additional information on pros and cons of different web-mapping libraries.

Google Maps JavaScript API[6] (Dincer and Uraz, 2013) is a proprietary web-mapping library by Google. The biggest advantage of the Google Maps API is that it brings the finely crafted look and feel of the Google Maps background layer to your own web map. On the other hand, background layers other than Google's are not supported. The Google Maps API also has advanced functionality not available elsewhere, such as integration with Street View[7] scenes. However the library is closed-source, which means the web maps cannot be fully customized. Also, it requires a paid subscription[8].

OpenLayers[9] (Gratier et al., 2015; Langley and Perez, 2016; Farkas, 2016) is an older, more mature, and more richly featured open-source JavaScript library for building web maps, otherwise very similar to Leaflet in its scope. However, OpenLayers is also more complex, heavier (in terms of JavaScript file size), and more difficult to learn. Leaflet can be viewed as a lighter and more focused alternative to OpenLayers.

ArcGIS API for JavaScript[10] (Rubalcava, 2015) is another commercial web-mapping solution. The ArcGIS API is primarily designed to be used with services published using ArcGIS Online[11] or ArcGIS Server[12], though general data sources can also be used. The ability of the ArcGIS API to tap into web services originating from ArcToolbox that perform geoprocessing on the server is a feature which has no good equivalent in open-source software. However, using an ArcGIS Server requires a paid (expensive) license. Also, the APIs are free to use for development or educational use, but require a fee for commercial use.

D3[13] (Murray, 2017) is an open-source JavaScript library for visualization[14] in general, though it is commonly used for mapping[15] too (Newton and Villarreal, 2014). D3 is primarily used for displaying vector layers, as raster tile layers are not well supported. D3 is probably the most complex and difficult to learn among the libraries listed in Table 6.1. However, it is very flexible and can be used to create truly innovative map designs. Go back to the examples from Section 0.1—some of the most impressive ones, such as *Earth Weather*[16], were created with the help of D3.

Mapbox GL JS[17] is a web-mapping library provided by a commercial company named Mapbox. Notably, Mapbox GL JS uses customizable vector tile layers (Section 6.5.6.3) as background. You can use existing basemaps, or build your own using an interactive "studio"[18] web application. Like Google Maps, Mapbox GL JS also requires a paid subscription[19], though the first 50,000 monthly map views are free.

---

[6]https://developers.google.com/maps/documentation/javascript/

[7]https://developers.google.com/maps/documentation/javascript/examples/streetview-service

[8]https://developers.google.com/maps/documentation/javascript/usage-and-billing

[9]https://openlayers.org/

[10]https://developers.arcgis.com/javascript/

[11]https://www.esri.com/software/arcgis/arcgisonline

[12]http://server.arcgis.com/en/server/latest/get-started/windows/what-is-arcgis-for-server-.htm

[13]https://d3js.org/

[14]https://bl.ocks.org/mbostock

[15]https://d3indepth.com/geographic/

[16]https://earth.nullschool.net/

[17]https://docs.mapbox.com/mapbox-gl-js/api/

[18]https://www.mapbox.com/mapbox-studio/

[19]https://www.mapbox.com/pricing/

6.5   Creating a basic web map

6.5.1   Overview

In this section, we will learn to create a basic web map using Leaflet. The map is going to contain a single background (tile) layer, initially zoomed-in on the Ben-Gurion University. The final result is shown in Figure 6.5.

6.5.2   Web page setup

We start with the following minimal HTML document, which we are familiar with from Chapter 1:

```
<!DOCTYPE html>
<html>
<head>
    <title>Basic map</title>
    <!-- More content will go here -->
</head>
<body>
    <!-- More content will go here -->
</body>
</html>
```

- Create an empty text file named `index.html` and copy the above code into that file.
- Follow the steps described in the Sections 6.5–6.8, to build your own web map from the ground up.

At this stage, the web page is empty. From here, we will do the following four things to add a map to our page:

- Include the Leaflet CSS file using a `<link>` element and the Leaflet JavaScript file using a `<script>` element (Section 6.5.3)
- Add a `<div>` element that will hold the interactive Leaflet map (Section 6.5.4)
- Add another, custom `<script>`, to create a map object and initiate the map inside the `<div>` element (Section 6.5.5)
- Add the tile OpenStreetMap basemap to our map using `L.tileLayer` (Sections 6.5.6–6.5.7)

6.5.3   Including Leaflet CSS and JavaScript

We need to include the Leaflet library on our web page before we can start using it. There are two options for doing this, just like we discussed when including the jQuery library (Section 4.5). We can either download the library files from the Leaflet website and load those local files, or we can use a hosted version of the files from a CDN. In the examples, we use the *local copy* option. Unlike jQuery, which consists of just one JavaScript file, Leaflet consists of two files: a JavaScript file and a CSS file. Leaflet also comes with associated image files that the code uses, such as the images used to display markers (Section 11.2.2).

To include the Leaflet CSS file, we add a `<link>` element referring to the file within the `<head>` section (after the `<title>`). We will use a local copy named `leaflet.css`:

```
<link rel="stylesheet" href="css/leaflet.css">
```

Remember that the path to the local file needs to correspond to the website directory structure (Section 5.5.2). For example, in the above `<link>` element, we are using a relative path: `css/leaflet.css`. This means that the `leaflet.css` file is located in a sub-directory named `css`, inside the directory where the HTML document is. For loading the file from a CDN (Section 4.5.3), we could replace the `css/leaflet.css` part with the following URL:

```
https://unpkg.com/leaflet@1.5.1/dist/leaflet.css
```

After including the Leaflet CSS file, we need to add a `<script>` element referring to the Leaflet JavaScript file. Again, we are going to load a local copy of the file, by placing the following element in the `<head>`:

```
<script src="js/leaflet.js"></script>
```

The path `js/leaflet.js` refers to a file named `leaflet.js` inside a sub-directory named `js`. For loading the Leaflet JavaScript file from a CDN, replace `js/leaflet.js` with the following URL:

```
https://unpkg.com/leaflet@1.5.1/dist/leaflet.js
```

Either way, after adding the above `<link>` and `<script>` elements, the Leaflet CSS and JavaScript files are linked to our web page, and we can begin working with the Leaflet library. Note that the local files provided in the online supplement, as well as the above remote URLs, refer to a specific version of Leaflet—namely Leaflet version `1.5.1`—which is the newest version at the time of writing (Table 0.2). In case we need to load a newer or older version, the local copies or URLs can be modified accordingly.

When working with a local copy of the Leaflet library, in addition to the JavaScript and CSS files we also need to create an `images` directory within the directory where our CSS file is (e.g., `css`), and place several PNG image files[20] there, such as `marker-icon.png` and `marker-shadow.png`. These files are necessary for displaying markers and other images on top of our map (Section 6.6.2). More information on how Leaflet actually uses those PNG images will be given in Section 11.2.2.

---

[20]See Appendix A for complete list of files and directory structure; see online version (Section 0.7) for downloading the complete set of files for the Leaflet library.

6.5.4   Adding map `<div>`

Our next step is to add a `<div>` element, inside the `<body>`. The `<div>` will be used to hold the interactive map. As we learned in Section 1.6.11, a `<div>` is a generic grouping element. It is used to collect elements into the same block group so that it can be referred to in CSS or JavaScript code. The `<div>` intended for our map is initially empty, but needs to have an ID. We will use JavaScript to "fill" this element with the interactive web map, later on:

```
<div id="map"></div>
```

In case we want the web map to cover the entire screen (e.g., Figure 6.5), which is what we will usually do in this book, we also need the following CSS rules:

```
body {
    padding: 0;
    margin: 0;
}
html, body, #map {
    height: 100%;
    width: 100%;
}
```

This CSS code can be added in the `<style>` element in the `<head>`. Recall that this method of adding CSS is known as embedded CSS (Section 2.7.3).

6.5.5   Creating a map object

Now that the Leaflet library is loaded, and the `<div>` element which will be used to contain it is defined and styled, we can move on to actually adding the map, using JavaScript. The Leaflet library defines a global object named L, which holds all of the Leaflet library functions and methods. This is conceptually very similar to the way that the jQuery library defines the $ global object. Using the `L.map` method, our first step is to create a map object. When creating a map object there are two important arguments supplied to `L.map`:

- The **ID** of the `<div>` element where the map goes in
- Additional map **options**, passed as an object

In our case, since the `<div>` intended for our map has `id="map"` (Section 6.5.4), we can initiate the map with the following expression, which goes inside a `<script>` element at the end of the `<body>`:

```
L.map("map");
```

As for additional map options[21], there are several ones that we can set. The options are passed together, as a single object, where property name refers to the option and the property value refers to the value which we want to set. The most essential options of `L.map` are those specifying the initially viewed map extent. One way to specify the initial extent is using

---

[21]You can always to refer to the documentation (`https://leafletjs.com/reference-1.5.0.html`) for the complete list of options for `L.map`, or any other Leaflet function.

the `center` and `zoom` option, passing the coordinates where the map is initially panned to and its initial zoom level, respectively. For example, to focus on the Ben-Gurion University we can indicate the `[31.262218, 34.801472]` location and set the zoom level to `17`. The `L.map` expression now takes the following form:

```
L.map("map", {center: [31.262218, 34.801472], zoom: 17});
```

Note that the `center` option is specified in **geographic coordinates** (longitude and latitude) using an array of the form `[lat, lon]` rather than `[lon, lat]`, that is, in the `[Y, X]` rather than the `[X, Y]` order. This may seem unintuitive to GIS users, but the `[lat, lon]` ordering is actually very common in many applications that are not specifically targeted to geographers, including Leaflet and other web-mapping libraries such as the Google Maps API[22]. When working with Leaflet, you need to be constantly aware of the convention to use `[lat, lon]` coordinates, to avoid errors such as displaying the wrong area (also see Section 7.3.2.1).

One more thing to keep in mind regarding coordinates is that most web-mapping libraries, including Leaflet, usually work with geographic coordinates (longitude, latitude) only as far as the *user* is concerned. This means that all placement settings (such as map center), as well as layer coordinates (Section 6.6), are passed to Leaflet functions as geographic coordinates (latitude, longitude), i.e., the **WGS84**[23] (`EPSG:4326`) coordinate reference system[24]. The map itself, however, is displayed using coordinates in a different system, called the **Web mercator**[25] projection (`EPSG:3857`) (Figure 6.1). For example, the following `[lon, lat]` coordinates (`EPSG:4326`) for central London:

```
[-0.09, 51.51]
```

will be internally converted to the following `[X, Y]` coordinates in Web mercator (`EPSG:3857`) before being displayed on screen:

```
[-10018.75, 6712008]
```

The user never has to deal with the Web mercator system, since it is only used *internally*, before drawing the final display on screen. However, it is important to be aware that it exists. As for the `zoom`, you may wonder what does the `17` zoom level mean. This will be explained when discussing tile layers (Section 6.5.6).

The `L.map` function returns a **map object**, which has useful methods and can be passed to methods of other objects (such as `.addTo`, see below). Therefore, we usually want to save that object in a variable, such as the one named `map` in the following expression, so that we can refer to it when running those methods later on in our script. Combining all of the above considerations, we should now have the following JavaScript code in the `<script>` element in our web page:

```
var map = L.map("map", {center: [31.262218, 34.801472], zoom: 17});
```

---

[22]https://developers.google.com/maps/documentation/javascript/adding-a-google-map
[23]https://en.wikipedia.org/wiki/World_Geodetic_System
[24]https://en.wikipedia.org/wiki/Spatial_reference_system
[25]https://en.wikipedia.org/wiki/Web_Mercator_projection

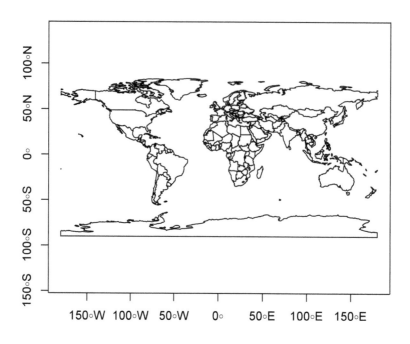

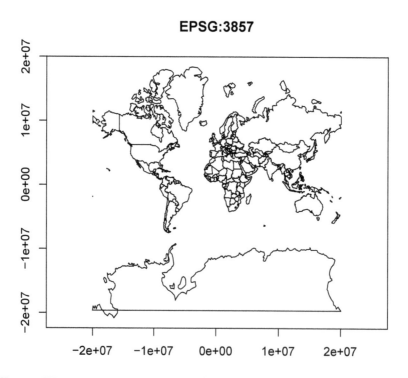

FIGURE 6.1: World map in the WGS84 (EPSG:4326) and Web mercator (EPSG:3857) projections

If you open the map at this stage, you should see that it has no content, just grey background with the "+" and "-" (zoom-in and zoom-out) buttons, which are part of the standard map interface. Our next step is to add a background layer (Section 6.2) on the map.

6.5.6 What are tile layers?

6.5.6.1 Overview

Tile layers are a fundamental technology behind web maps. They comprise the *background* layer in most web maps, thus helping the viewer to locate the *foreground* layers in geographical space. The word *tile* in tile layers comes from the fact that the layer is split into individual rectangular tiles. Tile layers come in two forms, which we are going to cover next: raster tiles (Section 6.5.6.2) and vector tiles (Section 6.5.6.3).

6.5.6.2 Raster tiles

The oldest and simplest tile layer type is where tiles are raster images, also known as **raster tiles**[26]. With raster tiles, tile layers are usually composed of PNG images. Traditionally, each PNG image is 256×256 pixels in size. A separate collection of tiles is required for each zoom level the map can be viewed on, with increasing numbers of tiles needed to cover a (global) extent in higher zoom levels. Conventionally, at each sequential zoom level, all of the tiles are divided into four "new" ones (Figure 6.2). For example, for covering the world at zoom level 0, we need just one tile. When we go to zoom level 1, that individual tile is split to $2 \times 2 = 4$ separate tiles. When we go further to zoom level 2, each of the four tiles is also split to four, so that we already have $4 \times 4 = 16$ separate tiles, and so on. In general, a global tile layer at zoom level z contains $2^z \times 2^z = 4^z$ tiles. At the default maximal zoom level in a Leaflet map (19), we need 274,877,906,944 tiles to cover the earth[27]!

The main rationale for tiled maps is that only the relevant content is loaded in each particular viewed extent. Namely, only those specific tiles that cover the particular extent we are looking at, in a particular zoom level, are transferred from the server. For example, just one or two dozen individual tiles (Figure 6.3) are typically needed to cover any given map extent, such as the one shown in Figure 6.5. This results in minimal size of transferred data, which makes tiled maps appear smooth and responsive.

The PNG images of all required tiles are sent from a **tile server**, which is simply a static server (Section 5.4.2) where all tile images are stored in a particular directory structure. To load a particular tile, we enter a URL such as `http://.../Z/X/Y.png`, where `http://.../` is a constant prefix, `Z` is the zoom level, `X` is the column (i.e., the x-axis) and `Y` is the row (i.e., the y-axis) (Figure 6.2). For example, the following URL refers to an individual tile at zoom level 17, column 78206 and row 53542, focused on building #72 at the Ben-Gurion University, from the standard OpenStreetMap tile server[28] (Figure 6.3).

`https://a.tile.openstreetmap.org/17/78206/53542.png`

Note how the URL for this tile is structured. The constant prefix of the URL is `https://a.tile.openstreetmap.org/`, referring to the OpenStreetMap tile server.

---

[26]`https://en.wikipedia.org/wiki/Tiled_web_map`
[27]See the *Zoom levels* tutorial (`https://leafletjs.com/examples/zoom-levels/`) for more information on tile layer zoom levels and their implementation in Leaflet.
[28]`https://wiki.openstreetmap.org/wiki/Tile_servers`

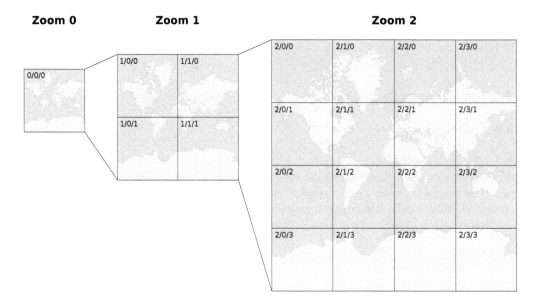

FIGURE 6.2: OpenStreetMap tiles for global coverage at zoom levels 0, 1, and 2. The Z/X/Y values (zoom/column/row) of each tile are shown in the top left corner.

The variable parts are specific values for Z (e.g., 17), X (e.g., 78206), and Y (e.g., 53542). You may recognize where this tile fits in the map view in Figure 6.5.

FIGURE 6.3: Individual OpenStreetMap raster tile, from zoom level 17, column 78206 and row 53542 (downloaded on 2019-03-29)

- Open the above URL in the browser to view the PNG image of the specific tile.
- Modify the Z, X, and Y parts of the URL to view other tiles.

6.5.6.3 Vector tiles

A more recent tile layer technology is where tiles are *vector* layers, rather than PNG images, referred to as **vector tiles**[29]. Vector tiles are distinguished by the ability to rotate the map while the labels keep their horizontal orientation, and by the ability to zoom in or out smoothly—without the strict division to discrete zoom levels that raster tile layers have (Figure 6.2). Major advantages of vector tiles are their smaller size and flexible styling. For example, Google Maps[30] made the transition from raster to vector tiles in 2013[31].

The Leaflet library does not natively support vector tiles, though there is a plugin called **Leaflet.VectorGrid**[32] for that. Therefore, in this book we will restrict ourselves to using raster tiles as background layers. There are other libraries specifically built around vector tile layers, such as the **Google Maps API** and **Mapbox GL JS**, which we mentioned previously (Section 6.4). The `example-06-01.html` shows a web map with a vector tile layer built with Mapbox GL JS (Figure 6.4). This is the only non-Leaflet web-map example that we are going to see in the book; it is provided for demonstration and comparison of raster and vector tiles.

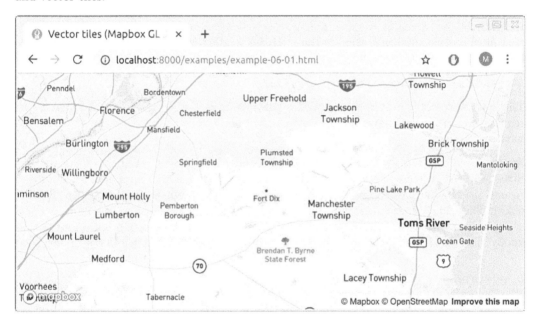

FIGURE 6.4: Screenshot of `example-06-01.html`

- Open `example-06-01.html` in the browser.
- Try changing map perspective by dragging the mouse while pressing the **Ctrl** key.
- Compare this type of interaction with that of `example-06-02.html`.
- What is the reason for the differences?

---

[29]https://en.wikipedia.org/wiki/Vector_tiles
[30]https://maps.google.com/
[31]https://en.wikipedia.org/wiki/Vector_tiles#Services_and_applications
[32]https://github.com/Leaflet/Leaflet.VectorGrid

6.5.7   Adding a tile layer

We now go back to discussing Leaflet and raster tile layers. Where can we get a tile layer from? There are many tile layers prepared and provided by several organizations, available on dedicated servers (Section 6.5.8) that you can add to your maps. Most of them are based on OpenStreetMap data (Section 13.2), because it is the most extensive free database of map data with global coverage. The tile layer we use in the following examples, and the one that the tile shown in Figure 6.3 comes from, is developed and maintained by OpenStreetMap itself. It is the default tile layer displayed on the `https://www.openstreetmap.org/` website.

To add a tile layer to a Leaflet web map, we use the `L.tileLayer` function. This function accepts:

- The **URL** of the tile server
- An object with additional **options**

Any raster tile server URL includes the `{z}`, `{x}`, and `{y}` placeholders, which are internally replaced by zoom level, column, and row each time the Leaflet library loads a given tile, as discussed previously (Section 6.5.6.2). The additional `{s}` placeholder refers to one of the available sub-domains (such as `a`, `b`, `c`), for making parallel requests to different servers hosting the same tile layer. Here is the URL for loading the default OpenStreetMap tile layer:

`https://{s}.tile.openstreetmap.org/{z}/{x}/{y}.png`

The options object is used to specify other parameters related to the tile layer, such as the required `attribution` (shown in the bottom-right corner on the map), or the `minZoom` and `maxZoom` levels[33].

For example, the following `L.tileLayer` call defines a tile layer using the default Open-StreetMap server and sets the appropriate attribution. The `.addTo` method is then applied to *add* the tile layer to the `map` object, referring to our web map defined earlier (Section 6.5.5). The `.addTo` method is applicable for adding any type of layer on our map, not just a tile layer, as we will see later on (Sections 6.6–6.8). In the attribution option, note that `&copy;` is a special HTML character entity to display the copyright symbol (©):

```
L.tileLayer(
    "https://{s}.tile.openstreetmap.org/{z}/{x}/{y}.png",
    {attribution: "&copy; OpenStreetMap"}
).addTo(map);
```

The attribution we used:

```
"&copy; OpenStreetMap"
```

can also be replaced with the following alternative version, where the word "OpenStreetMap" becomes a link (Section 1.6.8.1) to the OpenStreetMap Copyright and License[34] web page (see bottom-right corner in Figure 6.5):

---

[33]Again, for the complete list of options see the Leaflet documentation (`https://leafletjs.com/reference-1.5.0.html#tilelayer`).

[34]`https://www.openstreetmap.org/copyright`

```
'&copy; <a href="https://www.openstreetmap.org/copyright">OpenStreetMap</a>'
```

Now that the tile layer is in place, one last thing we are going to add to our first Leaflet map is the following `<meta>` element (Section 1.6.2.2) in the document `<head>`:

```
<meta name="viewport" content="width=device-width,
    initial-scale=1.0, maximum-scale=1.0, user-scalable=no">
```

This `<meta>` element disables unwanted scaling of the page when on mobile devices[35]. Without it, map symbols and controls will appear too small when viewed using a browser on a mobile device. Finally, our complete code for a basic Leaflet web map (`example-06-02.html`) is:

```
<!DOCTYPE html>
<html>
<head>
    <title>Basic map</title>
    <meta name="viewport" content="width=device-width,
        initial-scale=1.0, maximum-scale=1.0, user-scalable=no">
    <link rel="stylesheet" href="css/leaflet.css">
    <script src="js/leaflet.js"></script>
    <style>
        body {
            padding: 0;
            margin: 0;
        }
        html, body, #map {
            height: 100%;
            width: 100%;
        }
    </style>
</head>
<body>
    <div id="map"></div>
    <script>
        var map = L.map("map", {center: [31.262218, 34.801472], zoom: 17});
        L.tileLayer(
            "https://{s}.tile.openstreetmap.org/{z}/{x}/{y}.png",
            {attribution: '&copy; <a href="http://' +
            'www.openstreetmap.org/copyright">OpenStreetMap</a>'}
        ).addTo(map);
    </script>
</body>
</html>
```

The resulting map is shown in Figure 6.5.

---

[35] For more information, see the *Leaflet on Mobile* tutorial (`https://leafletjs.com/examples/mobile/`).

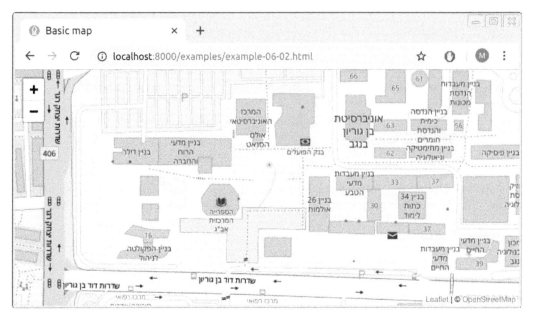

FIGURE 6.5: Screenshot of `example-06-02.html`

- Using the developer tools, we can actually observe the different PNG images of each tile being loaded while navigating on the map.
- Open `example-06-02.html` in the browser.
- Open the developer tools (in Chrome: by pressing **F12**).
- Go to the **Network** tab of the developer tools.
- Pan and/or zoom-in and/or zoom-out to change to a different viewed extent.
- In the Network tab, you should see a list of the new PNG images being transferred from the server, as new tiles are being shown on the map (Figure 6.6).
- Double-click on any of the PNG image file names to show the image itself.

6.5.8   Tile layer providers

As mentioned previously (Section 6.5.7), the default OpenStreetMap tile layer we just added on our map (Figure 6.5) is one of many available tile layers provided by different organizations. An interactive demonstration of various popular tile layer providers can be found in the *Leaflet Provider Demo*[36] web page. Once on the leaflet-providers web page, you can select a tile layer from the list in the right panel, then explore its preview in the central panel (Figure 6.7). Conveniently, the text box on the top of the page gives the exact `L.tileLayer` JavaScript expression that you can use in a `<script>` for loading the tile layer in a Leaflet map.

---

[36]`https://leaflet-extras.github.io/leaflet-providers/preview/`

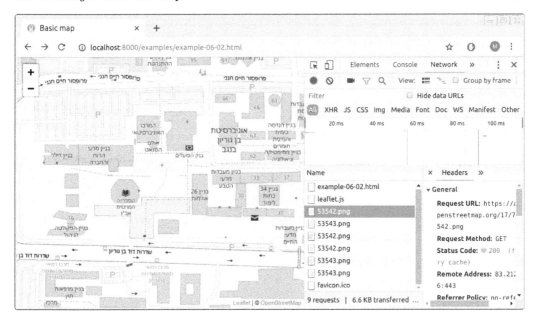

FIGURE 6.6: Observing network traffic as new tiles are being loaded into a Leaflet map

FIGURE 6.7: Interactive preview of tile layer providers

- Browse to the leaflet-providers page on `https://leaflet-extras.github.io/leaflet-providers/preview/` (Figure 6.7) and choose a tile layer you like.
- Replace the `L.tileLayer(...)` expression in the source code of `example-06-02.html` (after making a local copy) with the appropriate expression for your chosen tile layer.
- Save the file and refresh the page in the browser.
- You should now see the new tile layer you chose on the map!

6.6   Adding vector layers

6.6.1   Overview

So far we learned how to create a Leaflet map and add a tile layer on top of it. Tile layers are generic and thus typically serve as background, to help the user in positioning the foreground layers in geographical space. The foreground, on the other hand, is usually made of **vector layers**, such as points, lines, and polygons, though it is possible to add manually created image layers[37] too. There are several methods for adding vector layers on a Leaflet map.

In this section, we are going to add vector layers using those methods where the layer coordinates (i.e., its geometry) are manually specified by passing numeric arrays to the respective function. That way, we are going to add markers (Section 6.6.2), lines (Section 6.6.3), and polygons (Section 6.6.4) as foreground on top of the tile layer on our map. The latter coordinate-array methods are mostly useful for drawing simple, manually defined shapes. In Chapter 7, we will learn how to add vector layers based on GeoJSON strings rather than coordinate arrays, which is more useful to add pre-existing, complex layers.

6.6.2   Adding markers

There are three ways of *marking* a specific point on our Leaflet map:

- **Marker**—A PNG image, such as Leaflet's default blue marker (Figure 6.8)
- **Circle Marker**—A circle with a fixed radius in *pixels* (Figures 8.11, 12.6)
- **Circle**—A circle with a fixed radius in *meters*

To add a marker on our map, we use the `L.marker` function. The `L.marker` function creates a marker object, which we can add to our map with its `.addTo` method, just like we added the tile layer (Section 6.5.7). When creating the marker object with `L.marker` we specify the longitude and latitude where it should be placed, using an array of length 2 in the `[lat, lon]` format[38] (Section 6.5.5). For example, the following expression—when appended at the end of the `<script>` in `example-06-02.html`—adds a marker in the yard near building #72 at Ben-Gurion University:

---

[37] `https://leafletjs.com/reference-1.5.0.html#imageoverlay`

[38] Remember that the coordinates arrays that Leaflet accepts follow the `[lat, lon]` format, *not* the `[lon, lat]` format!

```
var pnt = L.marker([31.262218, 34.801472]).addTo(map);
```

You may wonder why we are assigning the marker layer to a variable, named `pnt` in this case, rather than just adding it to the map the way we did with the tile layer:

```
L.marker([31.262218, 34.801472]).addTo(map);
```

The answer is that the marker layer object contains useful methods, in addition to `.addTo`. We can apply those methods, later on, to make changes in the marker layer, such as adding popups on top of it (see Section 6.7 below). The resulting map `example-06-03.html`—with the additional marker as a result of including the above expression—is shown in Figure 6.8.

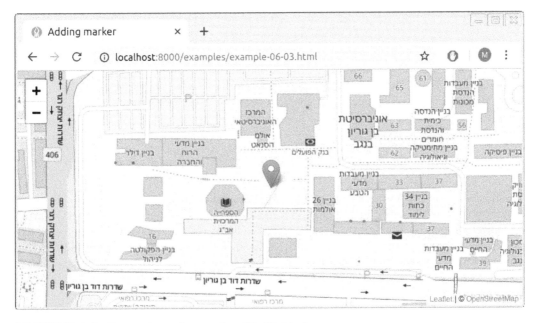

FIGURE 6.8: Screenshot of `example-06-03.html`

The marker is actually a PNG image, and you can replace the default blue marker with any other image, which we learn how to do in Section 11.2.2. However, we cannot otherwise easily control the size and color of the marker, since this would require preparing a different PNG image each time. Therefore, it is sometimes more appropriate to use *circle markers* or *circles* instead of markers.

With both circle markers and circles, instead of adding a PNG image, vector circles are being drawn around the specified location. You can set the size (radius) and appearance of those circles, by passing an additional object with options[39], such as `radius`, `color`, and `fillColor`. The difference between circle markers and circles is in the way that the `radius` is set: in pixels (circle marker) or in meters (circle). For example, the following expression adds a circle marker 0.001 degrees to the west from where we placed the ordinary marker:

---

[39]https://leafletjs.com/reference-1.5.0.html#circlemarker

```
L.circleMarker(
  [31.262218, 34.801472 - 0.001],
  {radius: 50, color: "black", fillColor: "red"}
).addTo(map);
```

- Run the above expression in the console inside `example-06-03.html`. A circle marker with the specified color scheme should appear on screen.
- Zoom in and out; the circle size should remain constant on screen, at 50 pixels.

Remember that in circle markers, created with `L.circleMarker`, the `radius` property is given in pixels (e.g., 50), which means the circle marker maintains constant size on screen, irrespective of zoom level. In circles, created using `L.circle`, the radius is set in meters, which means the circle size maintains constant spatial extent, expanding or shrinking as we zoom in or out.

6.6.3   Adding lines

To add a *line* on our map, we use the `L.polyline` function. Since a line is composed of several points, we specify the series of point coordinates the line goes through as an array of coordinate arrays. The internal arrays, i.e., the point coordinates, are specified just like marker coordinates, in the `[lat, lon]` format. In the following example, we are constructing a line that has just two points, but there could be many more points when making a more complex line. The line is drawn in the order given by the array, from the first point to the last one.

We can specify the appearance of the line, much like with the circle marker (Section 6.6.2), by setting various options such as line `color` and `weight` (i.e., width, in pixels). In case we are not passing any options, a default light blue 3px line will be drawn. In the following example, we override the default `color` and `weight` options, setting them to `"red"` and `10`, respectively:

```
var line = L.polyline(
  [[31.262705, 34.800514], [31.262053, 34.800782]],
  {color: "red", weight: 10}
).addTo(map);
```

Note that the list of relevant options differs between different layer types, so once more you are referred to the documentation[40] to check which properties may be modified for each particular layer type. For example, the `radius` option which we used for a circle marker is irrelevant for lines.

The resulting map `example-06-04.html`, now with both a marker and a line, is shown in Figure 6.9.

---

[40]https://leafletjs.com/reference-1.5.0.html#polyline

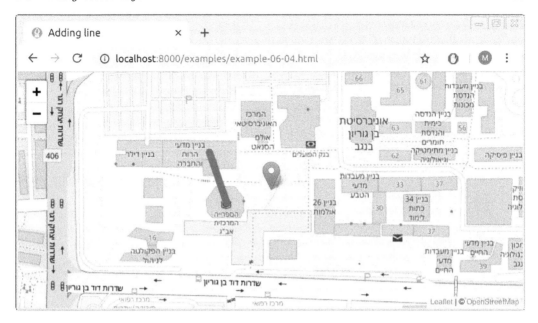

FIGURE 6.9: Screenshot of `example-06-04.html`

It is also possible to draw a **multi-part** line (Table 7.3) with `L.polyline`, by passing an array of separate line segments: *an array* of arrays of arrays. We will not elaborate on this option here, because this type of more complex multi-part shapes are usually loaded from existing GeoJSON layers (Chapter 7).

6.6.4 Adding polygons

Adding a *polygon* is very similar to adding a line, only that we use the `L.polygon` function instead of `L.polyline`. Like `L.polyline`, the `L.polygon` function also accepts an array of point coordinates the polygon consists of. Again, the polygon is drawn in the given order, from first node to last. Note that the array is not expected to have the last point repeating the first one, to "close" the shape (unlike in the GeoJSON format; see Section 7.3.2.2).

Like with circle markers, it is useful to define the border color and fill color of polygons, which can be done using the `color` and `fillColor` properties of the options object, respectively. For example, the following expression adds a polygon that has four nodes, with red border, yellow fill, and 4px border width:

```
var pol = L.polygon(
  [
    [31.263127, 34.803668],
    [31.262503, 34.803089],
    [31.261733, 34.803561],
    [31.262448, 34.804752]
  ],
  {color: "red", fillColor: "yellow", weight: 4}
).addTo(map);
```

The resulting map `example-06-05.html`, now with a marker, a line, and a polygon, is shown in Figure 6.10. Similarly to what we mentioned concerning lines (Section 6.6.3), it is also possible to use `L.polygon` to add more complex polygons such as **multi-part** polygons or polygons with **holes** (Table 7.3). As mentioned previously, we will use the GeoJSON format for adding such complex shapes (Chapter 7).

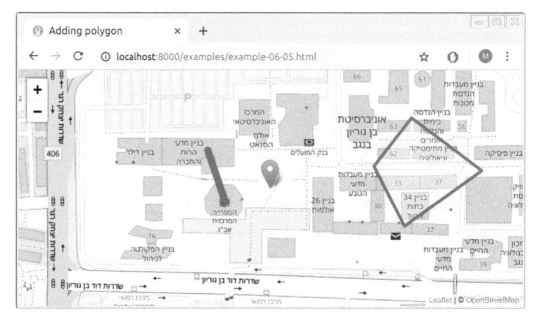

FIGURE 6.10: Screenshot of `example-06-05.html`

6.6.5   Other layer types

So far we mentioned most types of layers in Leaflet[41], as listed in the first six lines of Table 6.2. We haven't mentioned the rectangle layer, but this is practically a specific case of a polygon layer (Section 6.6.4) so there is not much to say about it. You are invited to try the `L.rectangle` example from Leaflet documentation on your own[42].

All of the latter layer types, except for tile layers, are vector layers drawn according to coordinate arrays. As mentioned above, the coordinate array method is useful for manually drawing simple shapes, but not very practical for complex vector layers, which are usually loaded from existing external sources. When working with complex predefined vector layers, the most useful type of Leaflet layer is probably `L.geoJSON`, specified using the GeoJSON format, which we learn about in Chapter 7.

Leaflet also supports adding **Web Map Service (WMS)**[43] layers using the `L.tileLayer.wms` function, which are (usually) raster images, dynamically generated by a WMS server (unlike tile layers which are pre-built). Finally, Leaflet supports **grouping** layers using functions called `L.layerGroup` and `L.featureGroup`. We will use these later on, in Chapters 7 and 10–13.

---

[41]The complete list of Leaflet layer types is given in the Leaflet documentation (`https://leafletjs.com/reference-1.5.0.html`).

[42]`https://leafletjs.com/reference-1.5.0.html#rectangle`

[43]`https://en.wikipedia.org/wiki/Web_Map_Service`

TABLE 6.2: Commonly used Leaflet layer types

Layer	Function
Tile layer	L.tileLayer
Marker	L.marker
CircleMarker	L.circleMarker
Circle	L.circle
Line	L.polyline
Polygon	L.polygon
Rectangle	L.rectangle
GeoJSON	L.geoJSON
Tile layer (WMS)	L.tileLayer.wms
Layer group	L.layerGroup
Feature group	L.featureGroup

6.7 Adding popups

Popups are informative messages which can be interactively opened and closed to get more information about the various features shown on a map. This is similar to the "identify" tool in GIS software. When the user clicks on a vector feature associated with a popup, an information box is displayed (Figure 6.11). When the user clicks on the "X" (close) symbol on the top-right corner of the box, on any other element on the map, or on the **Esc** key, the information box disappears.

Popups are added on the map by *binding* them to a given layer. Binding can be done using the `.bindPopup` method that any Leaflet layer object has. The `.bindPopup` method accepts a text string with the popup content. For example, we can add a popup to the line layer named **line** we created in Section 6.6.3, using the following expression:

```
line.bindPopup(
    "This is the path from <b>our department</b> to the <b>library</b>."
);
```

Note that popup content can contain HTML tags. In this case, we just use the `<b>` tag to specify bold font (Section 1.6.5), but you can use any HTML elements to add different type of content inside a popup, including links, lists, images, tables, videos, and so on. The resulting map with the popup is given in **example-06-06.html**[44]. Clicking on the line opens a popup, as shown in Figure 6.11.

---

[44]See the Leaflet *Quick Start Tutorial* (https://leafletjs.com/examples/quick-start/) for another overview on adding simple layers with popups in Leaflet.

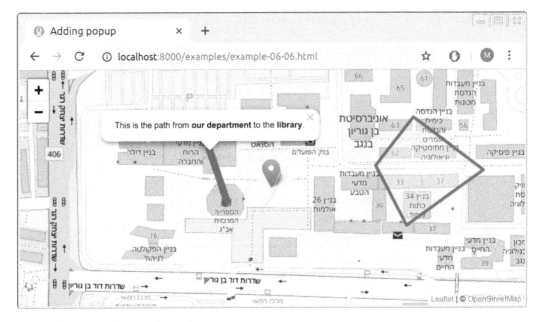

FIGURE 6.11: Screenshot of `example-06-06.html`

6.8   Adding a description

In Section 2.10 we saw an example of adding map title and description using custom HTML and CSS code (Figure 2.13). The Leaflet library has its own, simplified way to do the same task, using a function named `L.control`. With `L.control`, we can add custom map *controls* for different purposes, specified using custom HTML code. In the present example (`example-06-07.html`), we are going to use `L.control` to display a map description panel (Figure 6.13).

In `example-06-07.html`, we are using the `L.control` function to initialize a new map control named `legend`, and set its position to the bottom-left corner of the screen. The `L.control` function creates a new HTML element "above" the map, which can be filled with contents such as buttons or inputs (i.e., controls), map legends, or information boxes[45]:

```
var legend = L.control({position: "bottomleft"});
```

In the present example, we want to fill the control with a map description, composed of few lines of text and an image (Figure 6.12). We use the `.onAdd` method of the newly created control to define its contents and behavior. In this case, we are creating a `<div>` with HTML code which creates the map description:

---

[45]In subsequent chapters, we are going to use `L.control` to display other types of controls, such as a map legend (Section 8.6), a dynamically updated information box (Section 8.8.2), and a dropdown menu (Section 10.3).

```
legend.onAdd = function(map) {
    var div = L.DomUtil.create("div", "legend");
    div.innerHTML =
        '<p><b>Simple shapes in Leaflet</b></p><hr>' +
        '<p>This map shows an example of adding shapes ' +
        'on a Leaflet map</p>' +
        'The following shapes were added:<br>' +
        '<p><ul>' +
        '<li>A marker</li>' +
        '<li>A line</li>' +
        '<li>A polygon</li>' +
        '</ul></p>' +
        'The line layer has a <b>popup</b>. ' +
        'Click on the line to see it!<hr>' +
        'Created with the Leaflet library<br>' +
        '<img src="images/leaflet.png">';
    return div;
};
```

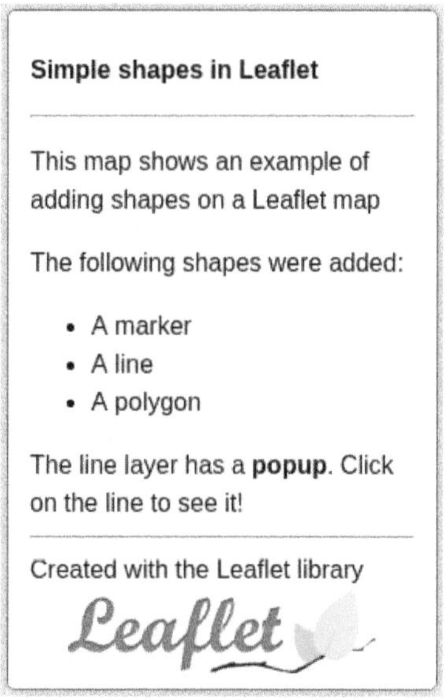

FIGURE 6.12: Map description

The above expression basically creates a `<div>` element with `id="legend"`, then adds internal HTML content into that element. Adding the HTML code is done with the plain JavaScript `.innerHTML` method, which we met in Section 4.3.3. We could do the same with jQuery (Section 4.7.2), but it would not make the code much simpler in this case. Note that we are using the `+` operators in order to split the assigned HTML code string into multiple lines and make the code more manageable and easier to read. Finally, the control is added to the map with `.addTo` method, just like we did for map layers in previous examples in this chapter:

```
legend.addTo(map);
```

The complete code for adding the map description is as follows:

```
var legend = L.control({position: "bottomleft"});
legend.onAdd = function(map) {
    var div = L.DomUtil.create("div", "legend");
    div.innerHTML =
        '<p><b>Simple shapes in Leaflet</b></p><hr>' +
        '<p>This map shows an example of adding shapes ' +
        'on a Leaflet map</p>' +
        'The following shapes were added:<br>' +
        '<p><ul>' +
        '<li>A marker</li>' +
        '<li>A line</li>' +
        '<li>A polygon</li>' +
        '</ul></p>' +
        'The line layer has a <b>popup</b>. ' +
        'Click on the line to see it!<hr>' +
        'Created with the Leaflet library<br>' +
        '<img src="images/leaflet.png">';
    return div;
};
legend.addTo(map);
```

We also need some CSS code to give our map description the final styling shown in Figure 6.12:

```
.legend {
    font-size: 16px;
    line-height: 24px;
    color: #333333;
    font-family: 'Open Sans', Helvetica, sans-serif;
    padding: 10px 14px;
    background-color: rgba(245,245,220,0.8) ;
    box-shadow: 0 0 15px rgba(0,0,0,0.2);
    border-radius: 5px;
    max-width: 250px;
    border: 1px solid grey;
}
.legend p {
    font-size: 16px;
    line-height: 24px;
}
.legend img {
    max-width: 200px;
    margin: auto;
    display: block;
}
```

Finally, we need to have the leaflet.png image file in the images sub-directory inside the directory where the HTML document is, for the last expression in our HTML code to work.

The expression adds the Leaflet library logo in the bottom of the map description (Figure 6.12):

```
<img src="images/leaflet.png">
```

The resulting map `example-06-07.html`, with the map description information box in the bottom-left corner, is shown in Figure 6.13.

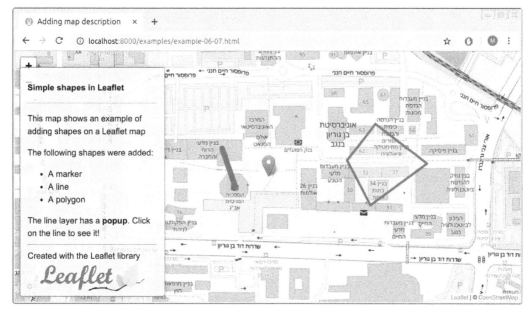

FIGURE 6.13: Screenshot of `example-06-07.html`

6.9 Introducing map events

In Section 4.3.4, we learned how the browser fires events in response to user interactions, and that these events can be handled using JavaScript to make our web page interactive. We also mentioned the event object (Section 4.11), which can be used to trigger specific responses according to event properties. Recall the mouse position in `example-04-07.html` (Figure 4.8), where mouse position was printed on screen thanks to the `pageX` and `pageY` properties of the event object. In Section 4.11, we also mentioned that *custom* event types and *custom* event object properties are often defined in JavaScript libraries for specific needs.

The Leaflet library defines specific event types and event properties appropriate for web mapping. These specific events can be captured for the entire map object, or for specific features the map contains. In this context, a popup is an example of a built-in event listener, with a pre-defined behavior: opening on click, closing when the "X" button or any other layer is clicked. Similarly, we can define and customize other Leaflet map interactive behaviors.

One of the most useful types of **map events** are `"click"` events, and the most important property of a map click event is the spatial location where the user clicked. When the user clicks on the map—in addition to the usual "click" event properties—the event object contains the latitude and longitude of where the click was made. The latitude and longitude are given in the special .latlng property of the map event[46] object. The .latlng event object property opens up the possibility of "capturing" the clicked coordinates and doing something useful with them.

For example, in Chapter 11 we will use the clicked location to query specific features—that are *near* the clicked location—from a database (Figure 11.8). In this section, we will see a simpler example, displaying the clicked coordinates inside a popup, which is opened on the clicked location itself (Figure 6.15). To display the currently clicked location using a popup, we need to do several things:

- Add an empty popup object to our map
- Write a function that, once called, sets popup location, fills the popup with content, and opens it on the map
- Add an event listener stating that each time the map is clicked, the latter function is executed

Starting with the first item, an empty popup object can be created with the L.popup function, as follows. This is in contrast to the .bindPopup method we used earlier to bind a popup to an *existing* layer (Section 6.7):

```
var popup = L.popup();
```

In the function that the event listener will refer to, hereby named onMapClick (see below), we will be using the event object parameter e. As mentioned above, when referring to map "click" events, the event object includes the .latlng property. The .latlng property is itself an object of type LatLng, which holds the longitude and latitude of the clicked location. This is basically a JavaScript object with two properties and two numeric values, such as the following one (plus some additional methods, which we don't need to worry about at the moment):

```
{lat: 31.2596049045682, lng: 34.80215549468995}
```

A LatLng object can be created with L.latLng, and it can be used for specifying coordinates instead of a simple [lat, lon] array. For example, try running the following expression in the console in any of the above examples. This will add a marker at a location defined using a LatLng object:

```
L.marker(L.latLng(31.264, 34.802)).addTo(map);
```

Back to the event listener example. The LatLng object, representing the clicked location, will be used for two purposes:

- Determining **where** the popup should be opened
- Setting the popup **contents**

These are accomplished with two methods of the popup object, .setLatLng and .setContent. Finally, once popup placement and content are set, the popup will be opened using a third

---

[46]https://leafletjs.com/reference-1.5.0.html#map-event

method named `.openOn`. The event listener function, hereby named `onMapClick`, thus takes the following form:

```
function onMapClick(e) {
    popup
        .setLatLng(e.latlng)
        .setContent(
            "You clicked the map at -<br>" +
            "<b>lon:</b> " + e.latlng.lng + "<br>" +
            "<b>lat:</b> " + e.latlng.lat
        )
        .openOn(map);
}
```

To better understand what this function does, try running the following code in the console in one of the Leaflet examples from this chapter:

```
var myLocation = L.latLng(31.264, 34.802);
L.popup()
    .setLatLng(myLocation)
    .setContent(
        "You clicked the map at -<br>" +
        "<b>lon:</b> " + myLocation.lng + "<br>" +
        "<b>lat:</b> " + myLocation.lat
    )
    .openOn(map);
```

This is basically the same code as the body of the `onMapClick` function, only that instead of `e.latlng` we used a specific `LatLng` instance, named `myLocation`. As a result of running this code section, you should see a popup with the content shown in Figure 6.14, opened at the location specified by `L.latLng(31.264, 34.802)`.

FIGURE 6.14: A popup displaying clicked location coordinates

Finally, we add an event listener specifying that the `onMapClick` function should be executed on map click. Note that in this case the element responding to the event is the entire `map`, so that any click inside the map triggers the `onMapClick` function:

```
map.on("click", onMapClick);
```

The complete code we need to add to a basic Leaflet map, so that clicked coordinates are displayed in a popup, is shown on the following page:

```
var popup = L.popup();
function onMapClick(e) {
    popup
        .setLatLng(e.latlng)
        .setContent(
            "You clicked the map at -<br>" +
            "<b>lon:</b> " + e.latlng.lng + "<br>" +
            "<b>lat:</b> " + e.latlng.lat
        )
        .openOn(map);
}
map.on("click", onMapClick);
```

The resulting map `example-06-08.html` is shown in Figure 6.15.

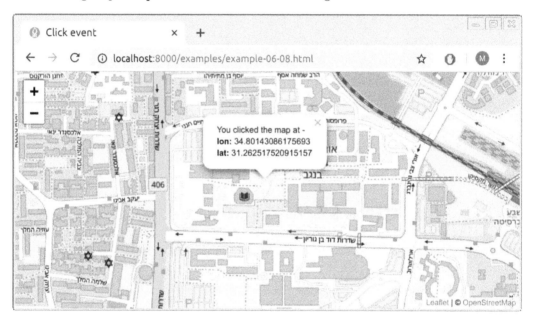

FIGURE 6.15: Screenshot of `example-06-08.html`

- Open `example-06-08.html` (Figure 6.15) and click anywhere on the map.
- As shown in Figure 6.15, the popup displays the `e.latlng` coordinates "as is", with many irrelevant digits.
- The first six digits refer to sub-meter (~0.1 m) precision, which is more than enough for most types of applications.
- Modify the `onMapClick` function so that the coordinates are rounded and only the first *six* digits of longitude and latitude are displayed inside the popup. (Hint: you can search on Google for a JavaScript function to round numbers according to specified number of digits.)

6.10   Exercise

- Create a Leaflet map with the location of places you want to visit, and the path you want to travel along between those places.
- The map should have the following components:
    - A tile layer
    - Markers of the visited locations
    - A line representing the travel path
    - Popups for each marker, containing location names
    - A description box, with your name and the list of locations in the right order
- Optional: Check out the *Layer Groups and Layers Control*[47] tutorial, and try to add a layer control for showing or hiding the markers and the line layers, and for switching between two or more types of tile layers (Figure 6.16).

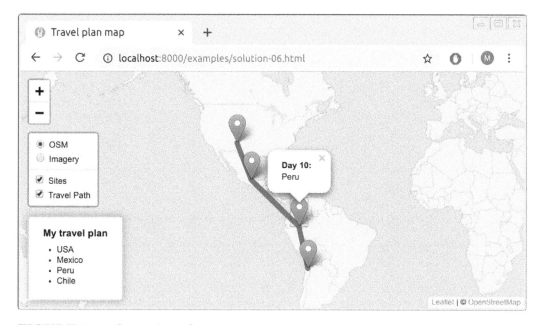

FIGURE 6.16: Screenshot of `solution-06.html`

---

[47]https://leafletjs.com/examples/layers-control/

7

GeoJSON

7.1 Introduction

In this chapter, we are going to learn about **GeoJSON**—a plain text format for vector layers. On the one hand, GeoJSON is a special case of JSON, which makes it fully compatible and easy to process with web technologies such as JavaScript. On the other hand, it is a fully-featured vector layer format, capable of representing complex vector layers and their non-spatial attributes. GeoJSON is, therefore, widely-used in web mapping and deserves its own chapter.

We are going to start with the definition of GeoJSON (Section 7.2) and an overview of its syntax and usage for representing various types of geometries (Sections 7.3–7.4). Then, we will discuss ways in which GeoJSON can be loaded and displayed on a Leaflet map (Sections 7.5–7.8).

7.2 What is GeoJSON?

GeoJSON is an plain-text format designed for representing vector geometries, with or without non-spatial attributes, based on the JavaScript Object Notation, JSON (Section 3.11.1). We briefly introduced the GeoJSON format in Section 3.11.2. This chapter is devoted to more in-depth treatment of the format and its use in web maps.

GeoJSON has become a very popular data format in many GIS technologies and services related to web mapping. It is actually the standard format for passing spatial vector layer data between the client and the server in web applications. As mentioned in Section 3.11.2, the main reason for its popularity is the fact that GeoJSON is a special case of JSON, which means that it can be easily parsed and processed with JavaScript. As a result, GeoJSON is the best (sometimes even the only) supported format by JavaScript web-mapping libraries and platforms, including those we learn about in this book: **Leaflet** (Chapters 6–8), **CARTO** (Chapters 9–11), and **Turf.js** (Chapter 12).

Other advantages of GeoJSON are that it is simple and human-readable, being a plain-text format. A disadvantage of GeoJSON is that its file size can get relatively large compared to other spatial vector layer formats, such as the **Shapefile**[1] or **GeoPackage**[2] formats.

---

[1]https://en.wikipedia.org/wiki/Shapefile
[2]https://en.wikipedia.org/wiki/GeoPackage

However, there are ways to reduce GeoJSON file size by simplifying its geometry and removing unnecessary attributes (Section 7.4.2).

In this chapter, we learn about how the GeoJSON format is structured (Section 7.3), how we can to create GeoJSON instances and edit them (Section 7.4), and how we can add GeoJSON layers on a Leaflet map (Sections 7.5–7.8). Keep in mind that, for simplicity, we will use the term GeoJSON interchangeably referring to both the original GeoJSON string, as well as the parsed object derived from that string in a JavaScript environment. Mapping library documentation often also uses the term GeoJSON for both cases. However, strictly speaking, the term GeoJSON refers just to text string instances, not to the derived JavaScript objects (see Section 3.11).

7.3   GeoJSON structure

7.3.1   Overview

In this section, we go over the structure of different types of GeoJSON strings you may encounter when working with this format. If you are new to JSON and GeoJSON, it may seem difficult to grasp the exact syntax of the GeoJSON format right from the start. Don't worry—we will come back to more examples later on in this chapter, as well as in the subsequent chapters. Moreover, rest assured one is almost never required to type GeoJSON strings *by hand*. Instead, web-map developers generally use pre-existing GeoJSON, exported from layers in other formats using GIS software, or coming from external web resources and databases. As we will see in the Section 7.4, you can create and edit GeoJSON even without GIS software, using web interfaces such as the one called **geojson.io**. Nevertheless, it is important to be familiar with the general structure of GeoJSON to recognize what type of layer you have at hand, and how to extract meaningful information from it.

GeoJSON is a format for representing **Simple Feature**[3] geometries, possibly along with their non-spatial attributes. The Simple Features standard defines 20+ types of geometry types, and the **Well-Known Text (WKT)**[4] format for representing them[5]. GeoJSON supports just the seven most commonly used Simple Feature geometry types (Figure 7.1). Having non-spatial attributes is not required for valid GeoJSON, so the seven geometries alone are encountered when representing geometric shapes only, with no attributes. In case non-spatial attributes *are* present, their combination with a geometry forms a `"Feature"`. Finally, a collection of more than one feature forms a `"FeatureCollection"`. The `"FeatureCollection"` GeoJSON type most closely corresponds to the meaning of a "layer", which you may be familiar with from GIS software (e.g., a Shapefile). The hierarchy of **GeoJSON types** thus includes nine types, which can be grouped into three "levels" of complexity:

- Geometry—One of seven Simple Feature geometry types, such as `"MultiPolygon"` (Table 7.1, Figure 7.1)
- `"Feature"`—A feature, i.e., a geometry along with its non-spatial attributes
- `"FeatureCollection"`—A collection of features

The seven **geometry types** that GeoJSON supports are listed in Table 7.1.

---

[3]https://en.wikipedia.org/wiki/Simple_Features
[4]https://en.wikipedia.org/wiki/Well-known_text
[5]We will come back to WKT in Section 9.6.3, when discussing spatial databases

TABLE 7.1: GeoJSON geometry types

Type	Description
`"Point"`	A single point
`"LineString"`	Sequence of connected points forming a line
`"Polygon"`	Sequence of connected points "closed" to form a polygon, possibly having one or more holes
`"MultiPoint"`	Set of points
`"MultiLineString"`	Set of lines
`"MultiPolygon"`	Set of polygons
`"GeometryCollection"`	Set of geometries of any type except for `"GeometryCollection"`

A `"Feature"` in GeoJSON contains a geometry object and additional non-spatial properties, also known as **attributes**. Finally, a `"FeatureCollection"` represents a collection of features. The following Sections 7.3.2–7.3.4 demonstrate the formatting of GeoJSON strings from the nine above-mentioned GeoJSON types:

- Section 7.3.2—7 geometry types
- Section 7.3.3—1 `"Feature"`
- Section 7.3.4—1 `"FeatureCollection"`

7.3.2 Geometries

7.3.2.1 General structure

A GeoJSON string for representing either of the first six geometry types, i.e., all types except for `"GeometryCollection"`, is composed of two properties[6]:

- `"type"`—The geometry type (a string, such as `"Point"`)
- `"coordinates"`—The coordinates (an array, such as `[30, 10]`)

For example, the following GeoJSON string is an example of a `"Point"` geometry:

```
{
  "type": "Point",
  "coordinates": [30, 10]
}
```

The `"type"` property can take one of the seven strings: `"Point"`, `"LineString"`, `"Polygon"`, `"MultiPoint"`, `"MultiLineString"`, `"MultiPolygon"` or `"GeometryCollection"` (Table 7.1). The `"coordinates"` property is specified with an array. The basic unit of the coordinate array is the **point coordinate**. According to the GeoJSON specification[7], point coordinates should refer to two-dimensional[8] locations in geographical units of longitude and latitude

---

[6]The seventh geometry type, `"GeometryCollection"`, has a slightly different structure which we discuss below (Section 7.3.2.4).

[7]https://tools.ietf.org/html/rfc7946

[8]Point coordinates in GeoJSON can also have three dimensions (3D) where the third dimension represents elevation, though this is less useful in web-mapping, and we will not encounter such examples in the book.

(ESPG:4326), i.e., [lon, lat][9]. The basic point coordinate can be used on its own (for "Point", as shown in the above GeoJSON example), or as a component in higher hierarchical level arrays for the other geometry types (see below).

Now that we covered the things that are common to the seven GeoJSON geometry types, we focus on the specifics of each. Overall, the seven geometry types can be conceptually divided into three groups, based on their complexity:

- "Point", "LineString", "Polygon"—**Single-part** geometries, where the geometry consists of *one* shape of one type (Section 7.3.2.2)
- "MultiPoint", "MultiLineString", "MultiPolygon"—**Multi-part** geometries, where the geometry consists of *one or more* shapes of one type (Section 7.3.2.3)
- "GeometryCollection"—**Geometry collections**, where the geometry can consist of one or more shapes of *any* type (Section 7.3.2.4)

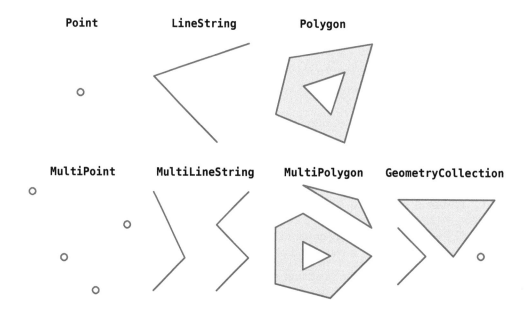

FIGURE 7.1: Seven Simple Feature geometry types supported by the GeoJSON format

7.3.2.2 Single-part geometries

For single-part geometry types, coordinates are specified as follows:

- "Point" coordinates are specified with a single point coordinate (e.g., [30, 10]).
- "LineString" coordinates are specified with an array of point coordinates (e.g., [[30, 10], [10, 30], [40, 40]]).
- "Polygon" coordinates are specified with an array of arrays of point coordinates, where each array of point coordinates specifies the **exterior** border ring (first array), or polygon **holes** (all other arrays, if any). Also, the last coordinate in each set is equal to the first one, to form a closed ring (e.g., [[[30, 10], [40, 40], [20, 40], [10, 20], [30, 10]]]).

---

[9]Remember that in Leaflet the convention is to specify point coordinates the *opposite* way, as [lat, lon] (Section 6.5.5). A nice blog post (https://macwright.org/lonlat/) by Tom MacWright gives a list of different mapping software using each of the [lon, lat] and [lat, lon] conventions.

Note how there is an increasing number of hierarchical levels in the coordinates array for these three geometry types:

- A "Point" has **one** level [...].
- A "LineString" has **two** levels [[...]].
- A "Polygon" has **three** levels [[[...]]].

Table 7.2 shows several examples of single-part geometry types, including two varieties of "Polygon" geometries: *with* holes and *without* holes.

TABLE 7.2: GeoJSON single-part geometries

Type	Example	
"Point"		```{`    "type": "Point",` "coordinates": [30, 10]`}```
"LineString"		```{`    "type": "LineString",` "coordinates": [`        [30, 10], [10, 30], [40, 40]`]`}```
"Polygon"		```{`    "type": "Polygon",` "coordinates": [`        [[30, 10], [40, 40], [20, 40], [10, 20],` [30, 10]]`    ]`}```
"Polygon"		```{`    "type": "Polygon",` "coordinates": [`        [[35, 10], [45, 45], [15, 40], [10, 20],` [35, 10]],`        [[20, 30], [35, 35], [30, 20], [20, 30]]`]`}```

7.3.2.3 Multi-part geometries

Multi-part geometry types are similar to their single-part counterparts. The only difference is that one more hierarchical level is added into the coordinates array, for specifying *multiple shapes*. Therefore:

- A "MultiPoint" has **two** levels [[...]].
- A "MultiLineString" has **three** levels [[[...]]].
- A "MultiPolygon" has **four** levels [[[[...]]]].

For example, a "MultiLineString" consists of an array of arrays of arrays, or an array of "LineString" coordinates, for defining several line **parts**, as in [[[10, 10], [20, 20], [10, 40]], [[40, 40], [30, 30], [40, 20], [30, 10]]]. Table 7.3 gives examples of multi-part geometry types.

TABLE 7.3: GeoJSON multi-part geometries

Type	Example

"MultiPoint"	`{`
	  `"type": "MultiPoint",`
	  `"coordinates": [`
	    `[10, 40], [40, 30], [20, 20], [30, 10]`
	  `]`
	`}`

"MultiLineString"	`{`
	  `"type": "MultiLineString",`
	  `"coordinates": [`
	    `[[10, 10], [20, 20], [10, 40]],`
	    `[[40, 40], [30, 30], [40, 20], [30, 10]]`
	  `]`
	`}`

"MultiPolygon"	`{`
	  `"type": "MultiPolygon",`
	  `"coordinates": [`
	    `[`
	      `[[30, 20], [45, 40], [10, 40], [30, 20]]`
	    `],`
	    `[`
	      `[[15, 5], [40, 10], [10, 20], [5, 10],`
	      `[15, 5]]`
	    `]`
	  `]`
	`}`

"MultiPolygon"	`{`
	  `"type": "MultiPolygon",`
	  `"coordinates": [`
	    `[`
	      `[[40, 40], [20, 45], [45, 30], [40, 40]]`
	    `],`
	    `[`
	      `[[20, 35], [10, 30], [10, 10], [30, 5],`
	      `[45, 20], [20, 35]],`
	      `[[30, 20], [20, 15], [20, 25], [30, 20]]`
	    `]`
	  `]`
	`}`

7.3.2.4 Geometry collections

A geometry collection is a set of several geometries, where each geometry is one of the previously listed six types, i.e., any geometry type excluding `"GeometryCollection"`. For example, a `"GeometryCollection"` consisting of two geometries, a `"Point"` and a `"MultiLineString"`, can be defined as follows:

```
{
  "type": "GeometryCollection",
  "geometries": [
    {
      "type": "Point",
      "coordinates": [...]
    },
    {
      "type": "MultiLineString",
      "coordinates": [...]
    }
  ]
}
```

where [...] are the coordinate arrays for each geometry. Table 7.4 shows an example of a geometry collection GeoJSON string.

TABLE 7.4: GeoJSON geometry collection

Type	Example

`"GeometryCollection"`	`{`
	`  "type": "GeometryCollection",`
	`  "geometries": [`
	`    {`
	`      "type": "Point",`
	`      "coordinates": [10, 30]`
	`    },`
	`    {`
	`      "type": "MultiLineString",`
	`      "coordinates": [`
	`        [[10, 10], [20, 20]],`
	`        [[40, 40], [30, 30], [40, 20],`
	`        [30, 10]]`
	`      ]`
	`    }`
	`  ]`
	`}`

It is harder to deal with geometry collections when applying spatial operations, because not every spatial operator (Section 9.6.4) has the same meaning (if any) for all geometry types. For example, calculating line length is meaningful for a `"MultiLineString"` geometry, but not for a `"Point"` geometry. For this reason, geometry collections are rarely encountered in practice, and we will not use them in this book.

7.3.3   Features

A "Feature" is formed when a geometry is combined with non-spatial attributes, to form a single object. The non-spatial attributes are encompassed in a property named "properties", containing one or more name-value pairs—one for each attribute. For example, the following "Feature" represents a geometry with two attributes, named "color" and "area":

```
{
  "type": "Feature",
  "geometry": {...},
  "properties": {
    "color": "red",
    "area": 3272386
  }
}
```

where {...} represents a geometry object, i.e., one of the seven geometry types shown above. Table 7.5 shows an example of a GeoJSON feature.

TABLE 7.5: GeoJSON feature

Type	Example
"Feature"	```{ "type": "Feature", "geometry": { "type": "Polygon", "coordinates": [[[15, 5], [40, 10], [10, 20], [5, 10], [15, 5]]] }, "properties": { "color": "red", "area": 3272386 } }```

7.3.4   Feature collections

A "FeatureCollection" is, like the name suggests, a collection of "Feature" objects. The separate features are contained in an array, comprising the "features" property. For example, a "FeatureCollection" composed of four features can be specified as follows:

```
{
  "type": "FeatureCollection",
  "features": [
    {...},
    {...},
```

```
      {...},
      {...}
   ]
}
```

where each `{...}` represents a `"Feature"`. Table 7.6 shows an example of a GeoJSON feature collection.

TABLE 7.6: GeoJSON feature collection

Type	Example
`"FeatureCollection"`	

```
{
    "type": "FeatureCollection",
    "features": [
      {
        "type": "Feature",
        "geometry": {
          "type": "Polygon",
          "coordinates": [
            [[30, 20], [45, 40], [10, 40],
            [30, 20]]
          ]
        },
        "properties": {
          "color": "green",
          "area": 3565747
        }
      },
      {
        "type": "Feature",
        "geometry": {
          "type": "Polygon",
          "coordinates": [
            [[15, 5], [40, 10], [10, 20],
            [5, 10], [15, 5]]
          ]
        },
        "properties": {
          "color": "red",
          "area": 3272386
        }
      }
    ]
}
```

If you are coming from GIS background, the `"FeatureCollection"` GeoJSON type will seem the most natural one. For example, in the terminology of the Shapefile—the most commonly used format for vector layers in GIS software—a `"FeatureCollection"` is analogous to a layer with more than one feature and one or more attributes—which is by far the most commonly encountered case. A "Feature" is analogous to a layer containing a single feature, which is rarely encountered. Bare GeoJSON geometries, such as "Point" or "MultiPolygon", have no analogs in the Shapefile format.

One more thing that may seem surprising for GIS software users is that a "FeatureCollection" does not have to be composed of features with the same type of geometry. For instance, in the above schematic GeoJSON "FeatureCollection", the first two {...} features may have "Point" geometry while the other two {...} features can have "Polygon" geometry (or any other geometry type combination). This kind of flexibility is also not supported in the Shapefile format, where all of the features must have the same geometry type[10].

7.4   Editing GeoJSON

7.4.1   geojson.io

The **geojson.io**[11] web application is great way to explore the GeoJSON format. This website contains an interactive map and a drawing control that you can use to draw new vector layers and to edit drawn ones. The GeoJSON string for the drawn content is displayed beside the map and is automatically synchronized with the currently drawn shapes while you are editing (Figure 7.2).

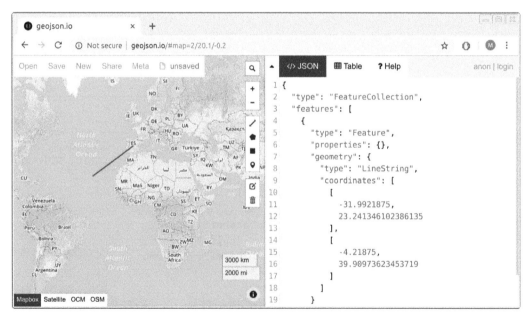

FIGURE 7.2: The geojson.io web application

---

[10]For more information on GeoJSON, the *More than you ever wanted to know about GeoJSON* blog post (https://macwright.org/2015/03/23/geojson-second-bite) is a recommended read. The complete specification of the GeoJSON format can be found on the official website (http://geojson.org/) of the standard.

[11]http://geojson.io

- Browse to `http://geojson.io`.
- Paste one of the GeoJSON string examples from Tables 7.2–7.6 into the right-side panel. You should see the shapes displayed on the left-side map panel.
- Use the drawing controls on the right of the map panel to draw some shapes on the map. You should see the GeoJSON string for the layer you created on the right-side text panel.

To understand a little better how web applications such as `http://geojson.io` work, we will shortly create our own (partial) version of such an application, using Leaflet and jQuery (Section 7.6).

7.4.2 mapshaper

In our introduction to the GeoJSON format (Section 7.2), we mentioned that a major disadvantage of this format is large file size. This is an especially painful limitation in web mapping, where we are limited by internet connection bandwidth and by browser processing capabilities. Displaying very large GeoJSON in a web map can result in bad user experience, since the web map will take a long time to load and will not be very responsive. Plainly speaking, GeoJSON size can get large when we have one or more of the following:

- A lot of **features**
- A lot of **attributes**
- High **precision** of the geometry (many digits in each coordinate)
- Highly **complex** geometry (many coordinates)

The first two are trivial to optimize: we just need to delete any non-essential features and attributes, keeping only the information actually displayed on the map. The third is also straightforward: we can round all coordinates to the minimal required precision, such as six digits for sub-meter (~0.1 m) precision (Section 6.9). The fourth—geometry complexity—is more tricky to deal with. We can't simply delete random coordinates of a complex line or polygon, because some coordinates are more important than others. Determining which coordinates should be kept and which ones can be safely deleted requires a **simplification** algorithm. For example, **Douglas–Peucker**[12] is a well-known geometry simplification algorithm, implemented in numerous GIS software.

mapshaper[13] is currently one of the best tools for geometry simplification. It is a free and open-source software with several geometry-editing functions, though it is best known for fast and easy simplification of vector layers. Importantly, mapshaper performs *topologically-aware* polygon simplification. This means that shared boundaries between adjacent polygons are always kept intact, with no gaps or overlaps, even at high levels of simplification. mapshaper comes in two versions:

- **Command-line tool**[14]
- **Interactive web interface** at `https://mapshaper.org/` (Figure 7.3)

---

[12]`https://en.wikipedia.org/wiki/Ramer%E2%80%93Douglas%E2%80%93Peucker_algorithm`
[13]`https://mapshaper.org/`
[14]`https://github.com/mbloch/mapshaper`

Take a moment to try mapshaper's web interface in the following exercise.

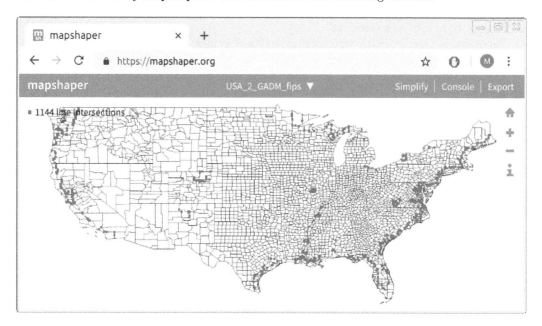

FIGURE 7.3: The `mapshaper.org` web application

- Download the file named `county2.geojson`, which is a detailed polygonal layer of U.S. counties, from the online version of the book (Appendix A), or obtain any other detailed polygonal layer from the web or from elsewhere. The `county2.geojson` file is very large (66.5 MB), and has some topological errors. We are going to fix the errors, then simplify the layer to optimize its usage in web maps, using **mapshaper**.
- Drag and drop the file into the `https://mapshaper.org/` main screen, or use the **Select** button, then click **Import**.
- The red dots you see (Figure 7.3) are line intersections, which are considered topological errors. To fix these, re-import the file, this time selecting the **snap vertices** option. There should be no red dots on the map now, since all line intersections have been fixed.
- Click the **simplify** button on the upper-right corner, then click **Apply**.
- Use the slider on the top of the page to select the level of simplification. Zoom-in on coastal areas with lots of details (such as in Florida), then move the slider to observe the simplification process more clearly.
- After choosing a level of simplification you are happy with, click **Export**, choose the GeoJSON format, then click **Export** once more. The simplified layer should now be downloaded. Check the size of the new file, which should be reduced according to the level of simplification you chose.

7.4.3 Formatting JSON

The GeoJSON examples in Section 7.3 are nicely formatted in a way that makes reading them easier. For example:

```
{
  "type": "Polygon",
  "coordinates": [
      [[35, 10], [45, 45], [15, 40], [10, 20], [35, 10]],
      [[20, 30], [35, 35], [30, 20], [20, 30]]
  ]
}
```

But what if you got the above GeoJSON from a database, where it is kept with no spaces or line breaks, like so:

```
{"type":"Polygon","coordinates":[[[35,10],[45,45],[15,40],...
```

This is a small example, but imagine a large `"FeatureCollection"`, with the entire GeoJSON string in a single line! There are numerous tools and web applications for automatically *formatting* JSON text strings that can be used in such a case. Formatting adds spaces and new lines so that it is once again easier to understand object structure. Here are two examples you can check out:

- `http://jsoneditoronline.org/`
- `https://jsonformatter.curiousconcept.com/`

Alternatively, you can always process the GeoJSON with `JSON.parse`, then examine the resulting object structure in the console, or produce a formatted string with `JSON.stringify(..., null, 4)` (Section 3.11.2).

7.5 Adding GeoJSON to Leaflet map

Below is an example of a GeoJSON string. This particular GeoJSON is a `"FeatureCollection"` (Section 7.3.4). It contains two features of type `"Polygon"`, representing the borders of two states in the U.S. You can tell the polygons are quadrilateral (i.e., have four edges) by the fact that each of them is defined with five coordinate pairs—recall that, in GeoJSON, the last polygon coordinate is equal to the first, to form a closed shape (Section 7.3.2.2). Each polygon has one attribute named `"party"`, with one of the polygons having the value `"Republican"` and the other polygon having the value `"Democrat"`.

```
{
  "type": "FeatureCollection",
  "features": [
    {
      "type": "Feature",
```

```
    "properties": {"party": "Republican"},
    "geometry": {
        "type": "Polygon",
        "coordinates": [
            [[-104.05, 48.99], [-97.22,  48.98],
            [-96.58,  45.94], [-104.03, 45.94],
            [-104.05, 48.99]]
        ]
    }
},
{
    "type": "Feature",
    "properties": {"party": "Democrat"},
    "geometry": {
        "type": "Polygon",
        "coordinates": [
            [[-109.05, 41.00], [-102.06, 40.99],
            [-102.03, 36.99], [-109.04, 36.99],
            [-109.05, 41.00]]
        ]
    }
}
]
}
```

- Open the console and define a variable named x with the above object.
- Type `x.features[0].geometry.coordinates[0][2][0]`.
- What is the meaning of the number you got?

How can we add the above GeoJSON layer to a Leaflet map? We will start with the basic map in `example-06-02.html` (Section 6.5) and make changes on top of it. Before we begin, let's focus our map on the U.S. area and reduce the zoom level, so that our polygons will be visible on the initial map extent:

```
var map = L.map("map").setView([43, -105], 4);
```

We are using the `.setView` method instead of the `L.map` option like we did in `example-06-02.html` (Section 6.5.7):

```
var map = L.map("map", {center: [43, -105], zoom: 4});
```

In the present example, both approaches are interchangeable. However, the `.setView` method offers more flexibility, since we can use it to dynamically modify our map-viewed extent, *after* the map was already created. Therefore it is useful to be familiar with both.

- Open the `example-07-01.html` (Figure 7.4) in the browser.
- Run the expression `map.setView([60, -80], 3)` in the console.
- What has happened?

Also, for the sake of diversity, let's replace the OpenStreetMap tile layer with a different one, called `CartoDB.PositronNoLabels`. The code shown below[15] was copied from the **Leaflet Provider Demo** website which was introduced in Section 6.5.8:

```
L.tileLayer("https://cartodb-basemaps-{s}.global.[...]/{z}/{x}/{y}.png", {
    attribution: '&copy; <a href="https://www.openstreetmap.org/[...]</a>',
    subdomains: "abcd",
    maxZoom: 19
}).addTo(map);
```

Now that we have a basic map with a tile layer focused on the U.S., let's add the GeoJSON layer. First, we will create a variable named **states**, and assign the above GeoJSON object to that variable[16]. Note that we are using the brackets notation {} to create the respective object out of the GeoJSON string right away, rather than entering it as text and using `JSON.parse` (Section 3.11.2):

```
var states = {
  "type": "FeatureCollection",
  ...
}
```

Second, we use the `L.geoJSON` function, to add a GeoJSON layer to our map. The function accepts a GeoJSON object, and transforms it to a Leaflet *layer* object. The layer can then be added on the map using its `.addTo` method, the same way we added tile layers, and simple point, line, and polygon shapes in Chapter 6. The expression to convert the GeoJSON object to a Leaflet layer and add it on the map is given below:

```
L.geoJSON(states).addTo(map);
```

After the last two expressions are added into our script, the resulting map (`example-07-01.html`) displays the GeoJSON polygons on top of the tile layer (Figure 7.4).

Note that the GeoJSON layer is added with the default style, as we did not pass any options to the `L.geoJSON` function. In Chapter 8, we will learn how to set custom GeoJSON style, either the same way for all features (Section 8.3), or differently depending on attributes (Section 8.4) or events (Section 8.8.1).

Also, keep in mind that, by default, Leaflet expects GeoJSON in **WGS84** (EPSG:4326) geographic coordinates, i.e., [`lon, lat`] (Section 6.5.5), which is what we indeed have in

---

[15]Parts of the URLs were replaced with [...] to save space.

[16]We omit most of the GeoJSON string to save space, since the complete string was already given above.

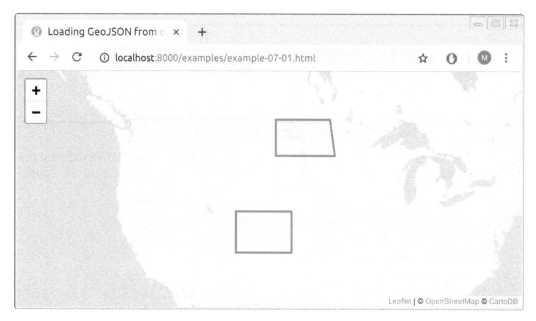

FIGURE 7.4: Screenshot of `example-07-01.html`

this example—as well as all other example in the book. As mentioned in Section 7.3.2.1, GeoJSON where the coordinates are given in other coordinate reference systems (CRS) does not conform to the GeoJSON specification, thus more rarely encountered and also less practical to use with Leaflet. For example, using non-WGS84 GeoJSON layers requires setting the entire map in a different CRS, which means that the standard tile layers cannot be loaded, as they are tailored for the WGS84 system. Unless there is some special reason to display the web map in a non-WGS84 CRS, in most cases it is more reasonable to just *transform* the GeoJSON to geographic coordinates before using it with Leaflet.

7.6 GeoJSON viewer example

7.6.1 Viewer structure

Earlier in this chapter, we used the **geojson.io** web application to interactively view and edit GeoJSON (Section 7.4.1). In this section, we are going to recreate a simplified GeoJSON viewer of our own. While doing it, we will learn some more about using GeoJSON with Leaflet. Like **geojson.io**, the interface in our GeoJSON viewer will have two parts:

- An interactive map
- A text editing area

Right below the text editing area, we are going to have a "submit" button (Figure 7.5). Pressing on the button will display the GeoJSON string currently typed into the text area on the interactive map.

7.6.2   HTML and CSS

Once again, we start with the basic map **example-06-02.html** (Figure 6.5), modifying and adding further components on top of it. First of all, instead of having a full screen map **<div id="map">**, we are going to have two **<div>** elements for the two parts of our page:

- **<div id="map">**—The map
- **<div id="text">**—The text input area

The HTML code of the two **<div>** elements is:

```
<div id="map"></div>
<div id="text">
    <textarea id="geojsontext"></textarea>
    <input type="button" id="submit" value="Submit">
</div>
```

Note that the second element (**<div id="text">**) contains two internal elements: a text input **<textarea>** (Section 1.6.12.5) and a button **<input type="button">** (Section 1.6.12.9). The **<textarea>** text input element is similar to the **<input type="text">** text input element (Section 1.6.12.4), but intended for *multi*-line rather than single-line text input. Next, we need some CSS to control the appearance and arrangement of these HTML elements:

```
#map {
    width: 60%;
    height: 100%;
    float: left;
}
#text {
    height: 100%;
    width: 40%;
    float: right;
}
#geojsontext {
    display: block;
    margin-left: auto;
    margin-right: auto;
    margin-top: 20px;
    width: 90%;
    height: 80%;
}
#submit {
    margin: 20px;
}
html, body {
    height: 100%;
    margin: 0;
    padding: 0;
}
```

First thing to note in the CSS code is that the map takes 60% of the screen width, while the text area takes 40% (Figure 7.5), as specified using the `width` properties. We are also using the `float` CSS property (which we haven't met so far), to specify that an element should be placed along the `left` or `right` side of its container. This places the map and the text entry `<div>` elements side-by-side, rather than one on top of the other.

We also need to load the jQuery library (Section 4.5), by placing the following element in the `<head>`:

```
<script src="js/jquery.js"></script>
```

The jQuery library is needed since we are going to use jQuery methods to get the currently entered text in the text area and to bind an event listener (Section 7.6.4 below).

7.6.3   Base map

Next, inside the `<script>`, we initialize a Leaflet map in the `<div id="map">` element, and add a tile layer, just like we did in `example-07-01.html` (Section 7.5)[17]:

```
var map = L.map("map").setView([0, 0], 1);
L.tileLayer("https://cartodb-basemaps-{s}.global.[...]/{z}/{x}/{y}.png", {
    attribution: '&copy; <a href="https://www.openstreetmap.org/[...]</a>',
    subdomains: "abcd",
    maxZoom: 19
}).addTo(map);
```

Note that we center the map to [0, 0] at zoom level 1. That way, the map initially displays a global extent (Figure 7.5).

7.6.4   Adding an event listener

All of the page elements are now in place: the map, the text area and the submit button. What's still missing in `example-07-02.html` to be functional, is the *association* between the text area and the map. To define it, we bind an event listener to the submit button.

The `showGeojson` function (see below) is going to collect the current value of the text area, and display the corresponding GeoJSON layer on the map with `L.geoJSON`. As we have seen in Section 7.5, `L.geoJSON` actually expects a parsed object rather than a text string. That is why the text string extracted from the `<textarea>` input needs to be parsed with `JSON.parse` before being passed to `L.geoJSON`. Here is the definition of the `showGeojson` function and the event listener:

```
function showGeojson() {
    var txt = $("#geojsontext").val();
    txt = JSON.parse(txt);
    L.geoJSON(txt).addTo(map);
}
$("#submit").on("click", showGeojson);
```

---

[17]Again, parts of the URLs were replaced with [...] to save space.

The event listener responds to "click" events on the "#submit" button. Each time the button is clicked, the current value of the "#geojsontext" text area is parsed and displayed on the map.

7.6.5 Using layer groups

One problem with the code that we have so far is that if we submit two (or more) different layers, they are sequentially added one on top of the other, on the same map. This can be inconvenient for the user, as previously entered layers will obstruct the new ones. We need some kind of mechanism to *remove* all previous layers before a new layer is loaded when pressing the submit button.

A convenient way of removing layers in Leaflet is to use **layers groups**. A layer group, in Leaflet terminology, is a collection of layers conveniently associated with a single variable. That way, we can apply the same action on *all* of the layers at once, as a single unit. An empty layer group can be created with `L.layerGroup()`, and added to the map with its `.addTo` method. For example, after initializing our map we can create a layer group named `layers`, and add it to our map, as follows:

```
var layers = L.layerGroup().addTo(map);
```

This expression, adding an empty layer group to a map, has no visible effect. However, any layer we add to that group, later on, will be automatically displayed on the map. At this stage, our map will contain a tile layer and an (empty) layer group. To add a GeoJSON layer to our layer group, and thus display it on the map, we simply replace `.addTo(map)` with `.addTo(layers)`, so that the following expression:

```
L.geoJSON(txt).addTo(map);
```

now becomes:

```
L.geoJSON(txt).addTo(layers);
```

The GeoJSON is displayed on the map in both cases. The advantage of the new approach, however, is that a layer group can be easily **cleared**, using the `.clearLayers` method[18]:

```
layers.clearLayers();
```

Clearing a layer group removes all layers that it previously contained, so that the layer group returns to its initial (empty) state. This means the layers are also removed from any map where the layer group was added on. Using the layer group approach, our modified `showGeojson` function is given below. Now, each time the submit button is pressed—all earlier GeoJSON content is removed from the map before the new content is shown:

---

[18]A layer group has another common use case: adding controls for toggling layer visibility (Figure 6.16). See the *Layer Groups and Layers Control* (`https://leafletjs.com/examples/layers-control/`) tutorial for an example.

```
function showGeojson() {
    layers.clearLayers();                    // Remove old GeoJSON
    var txt = $("#geojsontext").val();
    txt = JSON.parse(txt);
    L.geoJSON(txt).addTo(layers);            // Display new GeoJSON
}
```

The complete GeoJSON viewer application (`example-07-02.html`) is shown in Figure 7.5.

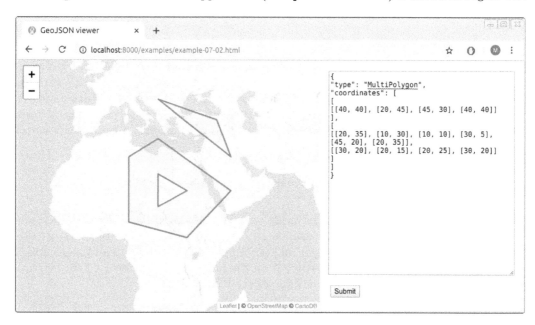

FIGURE 7.5: Screenshot of `example-07-02.html`

- Open `example-07-02.html` in the browser.
- Try copying and pasting some of the GeoJSON examples we saw earlier (Tables 7.2–7.6) to test our custom GeoJSON viewer!

7.7   Ajax

7.7.1   What is Ajax?

In the early days of the internet, up to mid-1990s, most websites were based on complete or *static* HTML pages served with a static web server (Section 5.4.2). Consequently, each user action required that a complete new page was loaded from the server. In many cases, this process is very inefficient and makes a bad user experience, as the entire page contents disappear and then the new page appears, even if just part of the page needs to be updated.

Nowadays, many modern websites use a set of techniques called **Asynchronous JavaScript and XML (Ajax)**[19] to *partially* update the contents of web pages. With Ajax, the web page can send data to, and retrieve data from, a server without interfering with the current state of the page or requiring page reload (Figure 7.6). That way, Ajax allows for web pages to change content *dynamically* without the need to reload the entire page. This makes the websites feel more responsive.

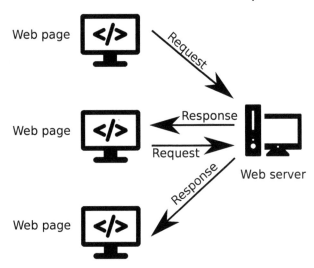

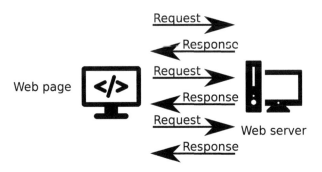

FIGURE 7.6: Schematic illustration of the difference between the traditional and Ajax request models

---

[19]https://en.wikipedia.org/wiki/Ajax_(programming)

Ajax uses an **asynchronous** processing model. This means the user can do other things while the web browser is waiting for the data to load, speeding up the user experience. The term asynchronous refers to the fact that loading data via Ajax does not stop the rest of the content from being loaded. Instead, the Ajax request is being sent to the server and in the interval before the server responds the rest of the page continues to load (Figure 7.7). When the response arrives, the content is processed. This is in contrast to the regular **synchronous** behavior of scripts, where the browser typically stops processing the page while executing each of the expressions in the `<script>`.

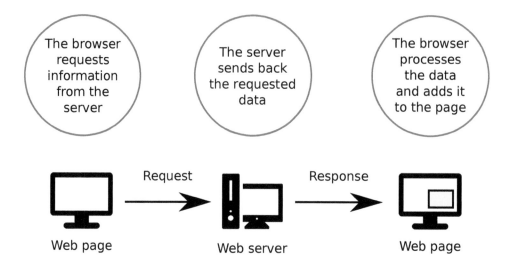

FIGURE 7.7: How Ajax works

XML or JSON are commonly used as the data exchange formats for communication between the server and the client using Ajax. In practice, most modern implementations use JSON, rather than XML, due to its advantages of being native to JavaScript. As we have seen in Section 3.11, JSON can be directly parsed to JavaScript objects, which makes processing of the data very convenient.

7.7.2 Ajax examples

You probably have seen Ajax used on many popular websites, even if you were not aware that it was being used. For example, the autocomplete feature in search boxes commonly uses Ajax. You have probably seen it used on the Google home page[20]. When you type into the search box on the Google, sometimes you will see results coming up before you have finished typing (Figure 7.8). What happens is that the currently entered text is sent to the server, using Ajax. The server then returns the relevant auto-complete suggestions. The suggestions are shown as a dropdown menu right below the search box.

Another example is when you scroll down the page on social network websites, such as Facebook[21], Instagram[22], or Twitter[23]. Once you reach the bottom of the page, *more* stories,

---

[20]https://www.google.com/
[21]https://www.facebook.com/
[22]https://www.instagram.com/
[23]https://twitter.com/

FIGURE 7.8: Auto-complete in Google search

images, or tweets are being loaded, and the page becomes longer. When you scroll down once more, again reaching the end of the page, more new content is loaded, and so on. In a way, the home page of those websites is nearly infinite. However, the entire page does not need to be loaded at once, nor does the page need to be completely reloaded each time new content is added at the bottom. When we reach the bottom of the page, new content is being requested using Ajax while all of the previous content and the navigation bars remain unmodified, which improves user experience.

Loading of tile layers on a Leaflet map (Section 6.5.7) is also an example of Ajax. The whole idea of tiles is based on the fact that only the relevant content, specific for the viewed extent and zoom level, is being loaded. Whenever we zoom in or out, or pan around the map, new tile PNG images are being requested using Ajax and added on the map (Figure 6.6), without ever reloading the entire web page. For example, if you're on a Leaflet map and scroll towards the north, the page's JavaScript sends an Ajax request the necessary new PNG images from the tile server. The server sends back its response—the new PNG images—which the JavaScript code then puts in the right place on the map[24].

7.7.3 Ajax requests with jQuery

Making Ajax requests is one more kind of task where jQuery greatly simplifies the usual JavaScript syntax, just like we previously saw regarding DOM queries (Sections 4.6–4.7) and iteration over objects (Section 4.12). jQuery provides several functions and methods that handle Ajax requests, summarized in Table 7.7.

---

[24]For a more detailed introduction to Ajax, also see the *Fetching Data From the Server* article (`https://developer.mozilla.org/en-US/docs/Learn/JavaScript/Client-side_web_APIs/Fetching_data`) by Mozilla.

TABLE 7.7: Methods for handling Ajax requests in jQuery

Method	Description
`.load()`	Loads HTML content into an existing HTML element
`$.get()`	Requests data from the server using HTTP `GET` request
`$.post()`	Sends data to be processed on the server using HTTP `POST` request
`$.getJSON()`	Loads and parses JSON data using HTTP `GET` request
`$.getScript()`	Loads and executes JavaScript code using HTTP `GET` request
`$.ajax()`	Performs customized Ajax request; all other methods use this method under the hood

Among these methods (Table 7.7), `.load` is special in that it is a method used on an existing HTML element, to add HTML content into it. The other five methods are general *functions* (i.e., methods of the `$` object). Also note that the first five functions and methods are in fact shortcuts for special cases of the sixth method `$.ajax`, which is the general function for making *any* kind of Ajax request.

For our purposes in this book, the most useful function is the `$.getJSON` function. The `$.getJSON` function can be used to load a JSON file using a `GET` request and immediately parse it to a JavaScript object. This is very convenient for loading and parsing GeoJSON strings, so that the GeoJSON layer can be immediately displayed on a web map. We are going to use the `$.getJSON` function to load GeoJSON layers on a Leaflet web map in most of the examples throughout the rest of the book. In Section 13.6, we will see an example of one more method from Table 7.7—the `$.post` function—to compose a `POST` request for sending data to the database in a crowdsourcing app.

7.8 The `$.getJSON` function

7.8.1 The `$.getJSON` function usage

To use the `$.getJSON` function, we first need to load the jQuery library. As we already know from Section 4.5, this can be done by adding the following `<script>` into the `<head>` of our document:

```
<script src="js/jquery.js"></script>
```

The basic usage of the `$.getJSON` function looks like this:

```
$.getJSON(url, callback);
```

where:

- `url`—The **URL** of the requested JSON file
- `callback`—The **function** to run if the request succeeds

The `callback` function has a parameter, e.g., named `data`, which refers to the **parsed object** returned from the server:

```
$.getJSON(url, function(data) {...});
```

For example, the following `$.getJSON` call uses an anonymous function that prints the parsed object `data`, obtained from the `url`, into the console:

```
$.getJSON(url, function(data) {
    console.log(data);
});
```

Of course, usually we want to do more than just printing with `console.log`. For example, the received object can be processed and used to append new HTML content on the page, display a new layer on a map, and so on. We will see examples of `$.getJSON` combined with various JSON and GeoJSON processing scenarios in subsequent chapters.

7.8.2 Loading local files

To add the contents of a **local** GeoJSON file as a layer on our map, we can set the `url` to a path of a GeoJSON file on our server. Then, we need to write a callback function (`function(data) {...}`) that executes once the GeoJSON is loaded and parsed. The callback function will contain the code that adds the GeoJSON to our map. To add GeoJSON to a Leaflet map, we use `L.geoJSON` (Section 7.5).

Let's try loading a sample GeoJSON file containing municipal boundaries of towns in Israel, named `towns.geojson`[25] and adding the layer on our map. The code for loading and displaying `towns.geojson` on the map is as follows:

```
$.getJSON("data/towns.geojson", function(data) {
    L.geoJSON(data).addTo(map);
});
```

Note that—in this particular example—the file `towns.geojson` is in the `data` sub-directory, relative to the HTML document location, therefore we are using the path `"data/towns.geojson"`. Also note that loading a local file with JavaScript through Ajax only works when viewing the page using a *web server*, such as Python's HTTP server, which was demonstrated in Section 5.6.2.4. This has to do with security restrictions placed by the browser: loading a local file via JavaScript is usually not permitted directly, but only through a server[26].

Figure 7.9 shows `example-07-03.html`, where the `towns.geojson` layer is loaded on a Leaflet map. Again, you may wonder how we can override the default style of the layer and set our own. As mentioned previously, this will be covered in Chapter 8.

---

[25] The `towns.geojson` file, like all files used in the examples, can be downloaded from the online version of the book (Section 0.7).

[26] There may be differences in this security restriction among different browsers. For example, at the time of writing, loading a local file is disabled in Chrome but works in Firefox (`https://stackoverflow.com/questions/38344612/ajax-request-to-local-file-system-not-working-in-chrome`).

FIGURE 7.9: Screenshot of `example-07-03.html`

7.8.3 Loading remote files

In the last example, `example-07-03.html` (Figure 7.9), we loaded a *local* GeoJSON file which was stored on the same server along with the HTML document. Using the same method, we can also load GeoJSON files stored in **remote** locations on the web[27].

For example, the United States Geological Survey (USGS) has a website dedicated to publishing earthquake location data in real time[28]. The website provides continuously updated records of recent earthquake locations. The data are given in several formats, including GeoJSON[29]. For example, the following URL leads to a GeoJSON file with the locations of earthquakes of magnitude above 4.5 in the past 7 days:

`https://earthquake.usgs.gov/earthquakes/feed/v1.0/summary/4.5_week.geojson`

We can replace `"data/towns.geojson"` from `example-07-03.html` with the above URL, thus loading remote earthquake locations layer instead of the local towns layer:

```
var url = "https://earthquake.usgs.gov/earthquakes/feed/v1.0/" +
    "summary/4.5_week.geojson";
$.getJSON(url, function(data) {
    L.geoJSON(data).addTo(map);
});
```

The resulting `example-07-04.html` is shown in Figure 7.10.

Note that due to security reasons, making Ajax requests from a *different domain* is not allowed by default. The mechanism defining this restriction is called **Cross-Origin Resource**

---

[27]Check out the *Leaflet GeoJSON* tutorial (`https://leafletjs.com/examples/geojson/`) for more details on loading GeoJSON.

[28]`https://earthquake.usgs.gov/earthquakes/feed/`

[29]`https://earthquake.usgs.gov/earthquakes/feed/v1.0/geojson.php`

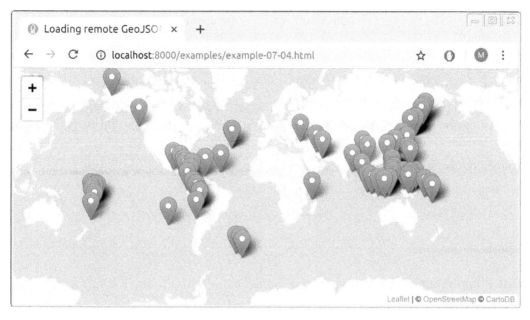

FIGURE 7.10: Screenshot of `example-07-04.html`

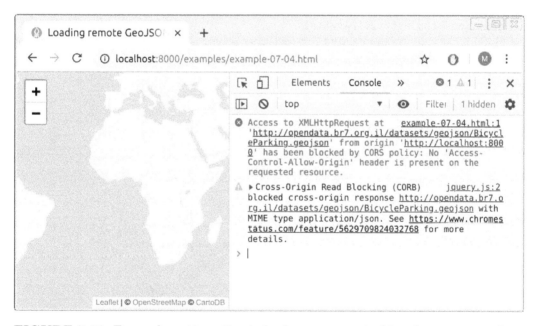

FIGURE 7.11: Error when attempting to load a resource, via Ajax, from a server where Cross-Origin Resource Sharing (CORS) is not allowed

Sharing (CORS)[30]. Basically, the server needs to allow remote connections from your specific domain (or from *any* domain) for the request to be successful. The USGS server, where the earthquakes GeoJSON is placed, allows CORS. This is why `example-07-04.html` works. In case we are trying to load a file from a server that *does not* allow CORS, the operation will fail displaying an error—such as the following one—in the JavaScript console (Figure 7.11):

---

[30]https://en.wikipedia.org/wiki/Cross-origin_resource_sharing

```
Failed to load http://opendata.br7.org.il/datasets/geojson/BicycleParking
.geojson:
No 'Access-Control-Allow-Origin' header is present on the requested
resource.
Origin 'http://localhost:8000' is therefore not allowed access.
```

If you encounter a CORS error while trying to load a resource, one option is to get the other server to allow CORS. If this is impossible, or you do not have access to the server, you can make the file available for your website through Ajax from a different server—or your own. CORS policy does not affect the access of server-side scripts to the resource, just the access of client-side JavaScript. Therefore an intermediate server, known as a **proxy server**[31], can be used to basically bypass the CORS restriction.

7.9 Exercise

- **Earth Observatory Natural Event Tracker (EONET)**[32] is a repository with real-time information about natural events. Conveniently, EONET publishes a real-time listing of natural events on Earth, available as a JSON file in `https://eonet.sci.gsfc.nasa.gov/api/v2.1/events`.
- Build a web map where locations of real-time *severe storms* are shown. The popup of each feature should display the storm name and observation time (Figure 7.12).

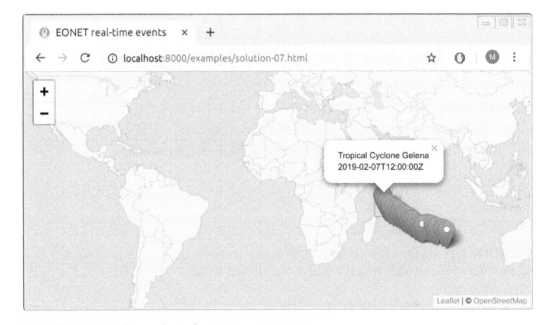

FIGURE 7.12: Screenshot of `solution-07.html`

---

[31]https://en.wikipedia.org/wiki/Proxy_server
[32]https://eonet.sci.gsfc.nasa.gov/

- Consider the following issues:
  - The JSON object has a property named `events`, which is an array of individual events.
  - Each event has an `id`, specifying event type, and `geometries`, which is an array of GeoJSON objects specifying event location and time.
  - To filter events of a specific type, use the categories specification given in `https: //eonet.sci.gsfc.nasa.gov/api/v2.1/categories`. For example, the categoty ID for `"Severe Storms"` event type is `10`.
- Hint: start with manually adding *one* geometry from the JSON file on the map, such as `data.events[0].geometries[0]`, where `data` is the parsed object. Then, try to generalize your code to iterate over all *events* and all *geometries* within each event.

8

Symbology and Interactivity

8.1 Introduction

Styling of map features lets us convey quantitative or qualitative *information* (how many residents are in that polygon?) or *emphasis* for drawing viewer attention (where is the border of the particular polygon of interest?). The way that aesthetic properties of map features are associated with underlying data or meaning is collectively known as map **symbology**, an essential concept of mapping in general, and web mapping in particular.

In Chapter 7, we learned about how GeoJSON layers can be added on a web map, whether from an object defined in the JavaScript environment (Section 7.5), a local file (Section 7.8.2), or a remote file (Section 7.8.3). As for style, however, all of the layers' features were drawn the same way. That is, because all features were set with the same default settings for their various aesthetic properties, such as fill color, line width, etc.

In addition to map symbology, web maps usually also express **interactive** behavior to further enhance user experience and convey even more information. Interactive behavior is what really sets interactive maps apart from static maps, such as those printed on paper or contained in image files and PDF documents. For example, an interactive map may have controls for turning overlaid layers on or off, switching between different base maps, displaying popups with textual or multimedia content for each clicked feature, highlighting elements on mouse hover, and so on.

In this chapter, we concentrate on defining map symbology (Sections 8.3–8.4) along with a corresponding legend (Section 8.6–8.7), by setting the style of our GeoJSON layers displayed on the map, and interactive behavior, by setting custom popups (Section 8.5), dynamic styling (Section 8.8.1), and dynamic information boxes (Section 8.8.2).

8.2 `L.geoJSON` options

As shown in Section 7.5, the `L.geoJSON` function accepts a GeoJSON object and turns it into a GeoJSON layer, which can be displayed on a Leaflet map. Additionally, the `L.geoJSON` function can also accept a number of options that we have not used yet. These options may be used to specify styling, filtering, attaching event listeners or popups, and otherwise controlling the display and functionality of the individual features comprising the GeoJSON layer. In this chapter, we use the following two options to enhance our GeoJSON layers:

- `style`—Determines layer **style**. The `style` can be either an *object* to style all features the same way (Section 8.3), or a *function* to style features based on their properties (Section 8.4).

- onEachFeature—A function that gets called on **each feature** before adding it to the GeoJSON layer. The onEachFeature function can be used to add specific popups (Section 8.5) or event listeners (Section 8.8.1) to each feature.

When using both options, the L.geoJSON function call may look like this:

```
L.geoJSON(geojson, {style: ..., onEachFeature: ...}).addTo(map);
```

where:

- geojson is the **GeoJSON** object; and
- {style: ..., onEachFeature: ...} is the **options** object, in this case including both style and onEachFeature options.

The ... parts in the above expression are to be replaced with an object or a function (see below), depending on how we want to define the appearance and behavior of the GeoJSON layer.

In addition to style and onEachFeature, in the exercise in the end of this chapter (Section 8.9), and two other examples later on (Sections 12.4.5 and 12.5.2), we will use a third option of L.geoJSON called pointToLayer. The pointToLayer option determines the way that GeoJSON point geometries are translated to visual layers. For example the pointToLayer option can be used to determine whether a GeoJSON "Point" feature should be displayed using a marker (Figure 7.10; the default), a circle marker (Figure 8.11), or a circle.

8.3   Constant style

The simplest way to style our GeoJSON layer is to set *constant* aesthetic properties for all of the features it contains. Just like in styling of line (Sections 6.6.3) and polygon (Section 6.6.4) layers, we only need to set those specific properties where we would like to override the default appearance. For example, let's start with example-07-01.html from Section 7.5. In that example, we used the following expression to add the GeoJSON object named states to our map:

```
L.geoJSON(states).addTo(map);
```

This sets the default Leaflet style for all polygons—blue border and semi-transparent blue fill (Figure 7.4). To override some of the default settings, the above expression can be replaced with the following one:

```
L.geoJSON(states, {
    style: {
        color: "red",
        weight: 5,
        fillColor: "yellow",
        fillOpacity: 0.2
    }
}).addTo(map);
```

In the new version, we are passing an object of options to the `L.geoJSON` function. The object contains one property named `style`, for which the value is also an object, containing the style specifications. In this example, we use four specifications:

- `color: "red"`—Border color = red
- `weight: 5`—Border width = 5px
- `fillColor: "yellow"`—Fill color = yellow
- `fillOpacity: 0.2`—Fill opacity = 20% opaque

The resulting map `example-08-01.html` is shown in Figure 8.1.

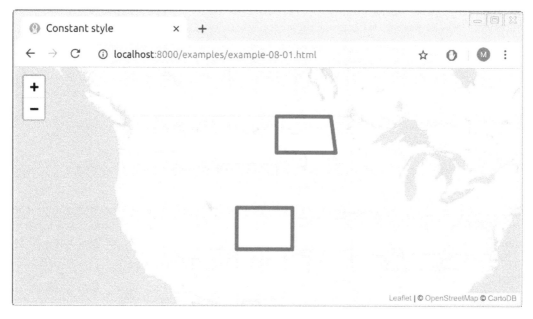

FIGURE 8.1: Screenshot of `example-08-01.html`

The options being set in this example—border width, border color, fill color, and opacity levels—are the most commonly used ones. Other options include line cap and join styles (`lineCap`, `lineJoin`) and line dash types (`dashArray`, `dashOffset`)[1].

8.4 Varying style

8.4.1 Setting varying style

In case we want the style to *vary* according to the properties of each feature, we need to pass a *function* instead of an *object* (Section 8.3), to the `style` option of `L.geoJSON`. We will demonstrate varying GeoJSON style through two examples: using the `states` GeoJSON object (Section 8.4.2) and the `towns.geojson` file (Section 8.4.3).

---

[1]The full list of styling options can be found in the *Path Options* (https://leafletjs.com/reference-1.5.0.html#path-option) section in the Leaflet documentation.

8.4.2 States example

The GeoJSON object named `states`, which we used in the last example (Section 8.3), has an attribute named `party` with two possible values: `"Republican"` and `"Democrat"`. To color the state polygons according to `party`, we need to write a function that sets feature style according to its properties. The function takes an argument `feature`, which refers the current GeoJSON feature, and returns the `style` object for that particular feature.

In the following example, the styling function, named `party_style`, returns an object with four properties. Three out of the four properties are set to constant values (`color`, `weight`, `fillOpacity`). The fourth property (`fillColor`) is variable. The `fillColor` property is dependent on the `party` attribute of the current feature (`feature.properties.party`):

```
function party_style(feature) {
    return {
        color: "black",
        weight: 1,
        fillColor: party_color(feature.properties.party),
        fillOpacity: 0.7
    };
}
```

The association between the `party` attribute and the fill color is made through the `party_color` function, separately defined below. The `party_color` function takes a party name `p` and returns a color name. The function uses `if`/`else` conditionals (Section 3.10.2):

```
function party_color(p) {
    if(p === "Republican") return "red"; else
    if(p === "Democrat") return "blue"; else
    return "grey";
}
```

The code of the `party_color` function specifies that if the value of `p` is `"Republican"`, the function returns `"red"`, and if the value of `p` is `"Democrat"`, the function returns `"blue"`. If any other value is encountered, the function returns the default color `"grey"`.

Finally, we can use the `party_style` function as the `style` option when creating the `states` layer and adding it on the map:

```
L.geoJSON(states, {style: party_style}).addTo(map);
```

The resulting map `example-08-02.html` is shown in Figure 8.2. Note that polygon fill color is now varying: one of the polygons is red while the other one is blue.

8.4.3 Towns example

As another example, we will modify the style of polygons in the Israel towns map, which we prepared in `example-07-03.html` (Figure 7.9). The towns layer in the `towns.geojson` file contains an attribute named `pop_2015`, which gives the population size estimate per town in

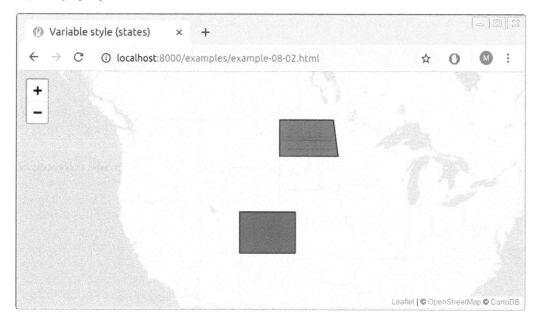

FIGURE 8.2: Screenshot of `example-08-02.html`

2015. We will use the `pop_2015` attribute to set polygon fill color reflecting the population size of each town (Figure 8.5).

First, we need to define a function that can accept an attribute value and return a color, just like the `party_color` function from the states example (Section 8.4.2). Since the `pop_2015` attribute is a continuous variable, our function will operate using **breaks** to divide the continuous distribution into categories. Each of the categories will be associated with a different color. The break points, along with the respective colors, comprise the layer *symbology*.

How can we choose the appropriate breaks when classifying a continuous distribution into a set of discrete categories? There are several commonly used methods for doing that. The simplest method, appropriate when the distribution is roughly uniform, is a **linear scale**. In a linear scale, break points are evenly distributed. For example, in case our values are between 0 and 1000, break points for a linear scale with five colors may be defined using the following array:

```
[200, 400, 600, 800]
```

Note that there are just three intervals between these values (200–400, 400–600, and 600–800), with the <200 and >800 ranges contributing the remaining two intervals. To emphasize this fact, we could explicitly specify the minimum and maximum values of the distribution (such as 0 and 1000), or just specify `-Infinity` and `Infinity`:

```
[-Infinity, 200, 400, 600, 800, Infinity]
```

When the distribution of the variable of interest is skewed, a linear scale may be inappropriate, as it would lead to some of the colors being much more common than others. As a result, the map may have little color diversity and appear uniformly colored. One solution is to use a **logarithmic scale**, such as:

```
[1, 10, 100, 1000]
```

The logarithmic scale gives increasingly detailed emphasis on one side of the distribution—the smaller values—by inducing increasingly wider categories. For example, the first category (1–10) is much narrower than the last category (100–1000).

A more flexible way of dealing with non-uniform distributions, which we are going to demonstrate in the present example, is using **quantiles**[2]. Quantiles are break points that distribute our set of observations into groups with equal counts of cases. For example, using quantiles we may divide our distribution into four equal groups, each containing a quarter of all values (0%–25%–50%–75%–100%)[3]. There are advantages and disadvantages[4] to using quantile breaks in mapping. The major advantage is that all colors are equally represented, which makes it easy to detect ordinal spatial patterns in our data. The downside is that category ranges can be widely different, in a non-systematic manner, which means that different color "steps" on the map are not consistently related to quantitative differences.

How can we calculate the quantile breaks for a given variable? There are many ways to calculate the break points in various software and programming languages, including in JavaScript[5]. One example of an external software to calculate quantiles is QGIS (QGIS Development Team, 2018). Once the layer of interest (such as `towns.geojson`) is loaded in QGIS, we need to set its symbology choosing the **Quantile (Equal Count)** for **Mode**, as shown in Figure 8.3. We also need to choose how many quantiles we want. For example, choosing 5 means that our values are divided into 5 equally sized groups, each containing 20% of the observations (i.e., 0%–20%–40%–60%–80%–100%). The breaks will be automatically calculated and displayed (Figure 8.3).

The four break points to split the `pop_2015` variable into five quantiles are given in the following array (Figure 8.3):

```
[399, 642, 933.2, 2089.8]
```

Once we figured out the break points, we can write a color-selection function, such as the one hereby named **getColor**. The function accepts a value **d** (population size) and determines the appropriate color for that value. Since we have five different categories, the function uses a hierarchical set of four conditionals, as follows:

```
function getColor(d) {
    if(d > 2089.8) return "..."; else
    if(d > 933.2) return "..."; else
    if(d > 642) return "..."; else
    if(d > 399) return "..."; else
        return "...";
}
```

---

[2]https://en.wikipedia.org/wiki/Quantile
[3]This specific case is called *quartiles* (https://en.wikipedia.org/wiki/Quartile).
[4]http://wiki.gis.com/wiki/index.php/Quantile
[5]https://simplestatistics.org/docs/#quantile

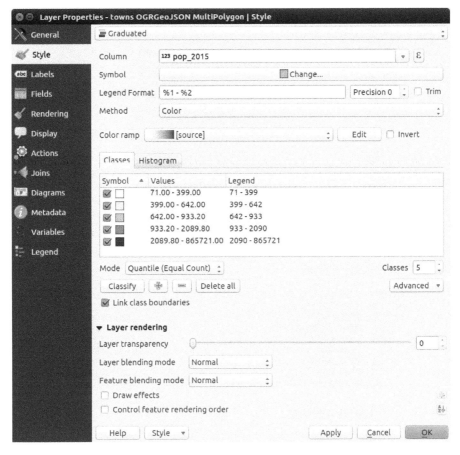

FIGURE 8.3: Setting symbology in QGIS, with automatically determined color scale breaks

What's missing in the body of the above function ("...") are the color definitions. As discussed in Section 2.8.2, there are several ways to specify colors with CSS. We can use **RBG** or **RGBA** values, **HEX** codes, or color **names** to specify colors in Leaflet. All of these methods are supported in Leaflet, too.

How can we pick the right set of five colors for our color scale? It may be tempting to pick the colors manually, but this may lead to an inconsistent color gradient. It is therefore best to use automatic tools and resources rather than try to pick the colors ourselves. There are many resources on the web and in software packages available for the job. One of the best and most well-known ones is **ColorBrewer**[6]. The ColorBrewer website provides a collection of carefully selected color scales, specifically designed for cartography by Cynthia Brewer[7] (Figure 8.4). Conveniently, you can choose the color palette, and how many colors to pick from it, then export the HEX codes of all colors in several formats, including a JavaScript array. ColorBrewer provides three types of scales:

- **Sequential** scales
- **Diverging** scales
- **Qualitative** scales

---

[6]http://colorbrewer2.org
[7]https://en.wikipedia.org/wiki/Cynthia_Brewer

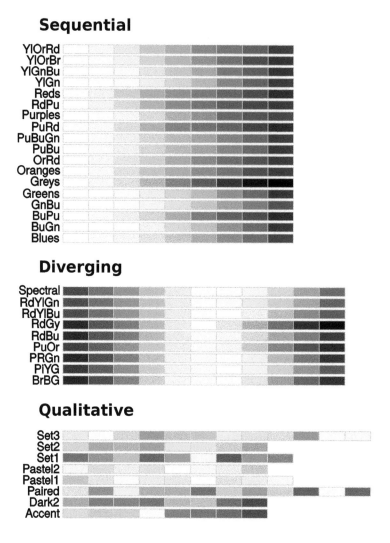

FIGURE 8.4: Sequential, diverging, and qualitative ColorBrewer scales, using the maximal number of colors available in each scale

For our population size data, we will use a **sequantial** scale (Figure 8.4), which is appropriate when the data are continuous and have a single direction (low-high). **Diverging** scales are appropriate for continuous variables that diverge in two directions around some kind of "middle" (low-mid-high). For example, rate of population growth can be mapped with a diverging scale, where positive and negative growth rates are shown with different increasingly dark colors (e.g., red and blue), while values around zero are shown with a light color (e.g., white). **Qualitative** scales are appropriate for categorical variables that have no inherent order, mapped to an easily distinguishable set of colors. For instance, land cover categories (e.g., built area, forests, water bodies, etc.) can be mapped to a qualitative scale[8].

---

[8]In a qualitative scale, it makes sense to choose intuitively interpretable colors whenever possible, e.g., built area = grey, water body = blue, forest = green, etc.

To export a JavaScript array of HEX color codes from ColorBrewer:

- Go to `http://colorbrewer2.org`.
- Select the **Number of data classes** (e.g., 5).
- Pick the **Nature of your data**, i.e., the scale type (e.g., "sequential").
- Select the scale in the **Pick a color scheme** menu (e.g., "OrRd" for an Orange-Red scale).
- Click **Export** and copy the contents of the text box named **JavaScript**.

For example, choosing five colors from the sequential color scheme called "OrRd" (Orange-Red), the ColorBrewer website gives us the following set of HEX color codes:

```
["#fef0d9", "#fdcc8a", "#fc8d59", "#e34a33", "#b30000"]
```

This is an expression for creating a JavaScript array, which we can copy and paste directly into our JavaScript code. The ColorBrewer website also contains a JavaScript code file named `colorbrewer.js`[9], with the complete set of color palette definitions as JavaScript arrays. If you use many different scales in a particular web map, it may be a good idea to include the file in your web page and then refer to the various color scale arrays by name in your JavaScript code.

Now that we have the five color codes, let's insert them into the `getColor` function definition. The function accepts a population size `d`, and returns the corresponding color code:

```
function getColor(d) {
    if(d > 2089.8) return "#b30000"; else
    if(d > 933.2) return "#e34a33"; else
    if(d > 642) return "#fc8d59"; else
    if(d > 399) return "#fdcc8a"; else
        return "#fef0d9";
}
```

One disadvantage of the above `getColor` function definition is that it uses a rigid set of `if`/`else` conditionals, which will be inconvenient to modify in case we decide to change the *number* of break points and colors. A more general version of the function could operate on an array of break points and a corresponding array of colors. Then, instead of `if`/`else` conditionals the function could use a `for` loop (Section 3.10.3.1). The loop goes over the breaks array, detects between which pair of breaks our value is situated, and returns the appropriate color. Here is an alternative version of the `getColor` function, using a `for` loop instead of `if`/`else` conditionals:

```
var breaks = [-Infinity, 399, 642, 933.2, 2089.8, Infinity];
var colors = ["#fef0d9", "#fdcc8a", "#fc8d59", "#e34a33", "#b30000"];

function getColor(d) {
    for(var i = 0; i < breaks.length; i++) {
        if(d > breaks[i] && d <= breaks[i+1]) {
            return colors[i];
        }
    }
}
```

---

[9]`http://colorbrewer2.org/export/colorbrewer.js`

With this version, it is easier to replace the color scale whenever necessary. All we need to do is modify the **breaks** and **colors** arrays.

As shown previously (**example-08-02.html**), our next step is to wrap the **getColor** function into another function, hereby named **style**, which is responsible for setting all of the styling options where we override the defaults:

```
function style(feature) {
    return {
        fillColor: getColor(feature.properties.pop_2015),
        weight: 0.5,
        opacity: 1,
        color: "black",
        fillOpacity: 0.7
    };
}
```

This time, we have four *constant* properties (**weight**, **opacity**, **color**, and **fillOpacity**) and one *variable* property (**fillColor**). Finally, we need to pass the **style** function to the GeoJSON **style** option when creating the towns layer and adding it on the map:

```
$.getJSON("data/towns.geojson", function(data) {
    L.geoJSON(data, {style: style}).addTo(map);
});
```

The resulting map **example-08-03.html** is shown in Figure 8.5. The towns polygons are now filled with one of the five colors from the Orange-Red ColorBrewer scale, according to the town popupation size, as defined in the **getColor** function.

FIGURE 8.5: Screenshot of **example-08-03.html**

There is a lot of theory and considerations behind choosing a **color scale** for a map, which we barely scratched the surface of. While we mainly focused on the technical process of defining and applying a color scale in a web map, the reader is referred to textbooks on cartography (Dent et al., 2008) and data visualization (Tufte, 2001; Wilke, 2019) for more information on the theoretical considerations of choosing a color scale for particular types of data and display purposes.

8.5 Constructing popups from data

In Section 6.7, we introduced the `.bindPopup` method for adding popups to simple shapes, such as the popup for the line between the Aranne Library and the Geography Department in `example-06-06.html` (Figure 6.11). We could do the same with a with a GeoJSON layer to bind the same popup to all features. However, it usually doesn't make much sense for all features (e.g., town polygons) to share the same popup. Instead, we usually want to add *specific* popups per feature, where each popup conveys information about the respective feature where it is binded. For example, it makes sense for the popup of each town polygon to display the name and population size of that specific town.

To bind specific popups, we use another option of the `L.geoJSON` options object, called `onEachFeature`. The `onEachFeature` option applies a function on each feature when that feature is added to the map. We can use the function to add a popup with specific content, based on the feature properties. The function we supply to `onEachFeature` has two parameters:

- `feature`—Referring to the current **feature** of the GeoJSON object being processed
- `layer`—Referring to the **layer** being added on the map

For example, the `onEachFeature` function can utilize `feature.properties` to access the properties of the currently processed feature, then run the `layer.bindPopup` method to add a corresponding popup for that specific feature on the map. The code below uses the `onEachFeature` option when loading the `towns.geojson` file, executing a function that binds a specific popup to each GeoJSON feature:

```
$.getJSON("data/towns.geojson", function(data) {
    geojson = L.geoJSON(data, {
        onEachFeature: function(feature, layer) {
            layer.bindPopup(feature.properties.name_eng);
        },
        style: style
    }).addTo(map);
});
```

Note how the `L.geoJSON` function now gets an options object with two properties: the `onEachFeature` property contains an anonymous function defined inside the object, while the `style` property contains a named function, incidentally having the same name `style`, defined above (Section 8.4.3).

One more thing introduced in this code section is that the *reference* to the GeoJSON layer is assigned to a variable, hereby named `geojson`. This will be useful in subsequent examples, where we will use the reference to the GeoJSON layer to execute its methods (Sections 8.8.1 and 8.8.3). Since we are assigning a value to the `geojson` variable, we should define it with an expression such as the following one, at the beginning of our script:

```
var geojson;
```

Also note that the previous code section binds a simple popup, with just the town name, using the contents of the **name_eng** property. However, as we have seen in Section 6.7, we can construct more complicated popup contents by concatenating several strings along with HTML tags. We can also combine several feature properties in the popup contents, rather than just one. For example, we can put both the town name and the town population size inside the popup of each town, and place them in two separate lines. To do that, we replace the expression shown previously:

```
layer.bindPopup(feature.properties.name_eng);
```

with this expanded one:

```
layer.bindPopup(
    '<div class="popup">' +
    feature.properties.name_eng + '<br>' +
    '<b>' + feature.properties.pop_2015 + '</b>' +
    '</div>'
);
```

Now the popups will show both the town name (**name_eng**) and the town population size (**pop_2015**). The latter is shown in bold font using the **<b>** element. Additionally, the entire popup contents are encompassed in a **<div>** element with **class="popup"**. The reason we do that is so that we can apply CSS rules for styling the popup contents. In this example, we just use the **text-align** property (Section 2.8.3.3) to make the popup text centered:

```
.popup {
    text-align: center;
}
```

The resulting map **example-08-04.html**, now with both variable styling and specific popups per feature, is shown in Figure 8.6. The popup for the Tel-Aviv polygon was opened for demonstration.

- It is more convenient to read long numbers when they are formatted with commas. For example, 432,892 is easier to read than 432892.
- Use a JavaScript function, such as the following one taken from a **StackOverflow**[10] question, to format the **pop_2015** values before including them in the popup.

---

[10]https://stackoverflow.com/questions/2901102/how-to-print-a-number-with-commas-as-thousands-separators-in-javascript

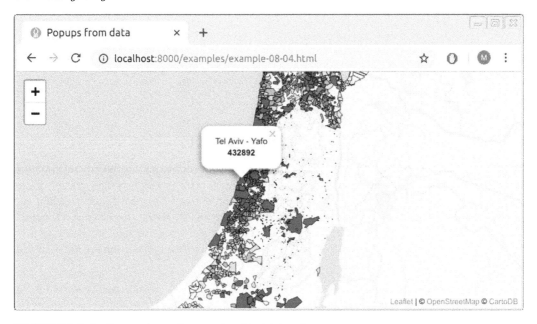

FIGURE 8.6: Screenshot of `example-08-04.html`

```
function formatNumber(num) {
    return num
        .toString()
        .replace(/(\d)(?=(\d{3})+(?!\d))/g, "$1,");
}
```

8.6 Adding a legend

In Section 6.8, we used the `L.control` function to add a map description. In this section, we use the same technique to create a **legend** for our map. The workflow for creating a legend involves creating a *custom control* with `L.control`, populating it with HTML that represents the legend components, then styling it with CSS so that the contents appear properly on screen. The following code section does all that, adding a legend to the towns population map from the last example (`example-08-04.html`):

```
var legend = L.control({position: "topright"});
legend.onAdd = function(map) {
    var div = L.DomUtil.create("div", "legend");
    div.innerHTML =
        '<b>Population in 2015</b><br>by Town<br>' +
        '<small>Persons/Town</small><br>' +
        '<i style="background-color: #b30000"></i>2090+<br>' +
```

```
          '<i style="background-color: #e34a33"></i>933 - 2090<br>' +
          '<i style="background-color: #fc8d59"></i>642 - 933<br>' +
          '<i style="background-color: #fdcc8a"></i>399 - 642<br>' +
          '<i style="background-color: #fef0d9"></i>0 - 399<br>';
     return div;
};
legend.addTo(map);
```

So, what did we do here? First, we created an instance of a custom control object, naming it `legend`. We used the `position` option to locate the control in the top-right corner of our map. Next, we used the `.onAdd` method of the `legend` control to run a function when the legend is added. The function creates a new `<div>` in the DOM, giving it a class of `"legend"`. This will let us use CSS to style the entire legend using the `.legend` selector (Section 8.7). We then populate the newly created `<div>` with HTML by using the `.innerHTML` method, like we already did in Section 6.8.

Most of the HTML code should be familiar from Chapter 1. One element type which we have not seen yet is `<small>`, used to create relatively smaller text, which is convenient for displaying the units of measurement (Figure 8.7):

```
<small>Persons/Town</small>
```

What may seem strange is that we use the `<i>` (italic text) to represent our legend **symbols**. The `<i>` elements are useful thanks to the fact they are colored using inline CSS (Section 2.7.2) and the `background-color` property (Section 2.8.2.1). The five `<i>` elements thus reflect the colors corresponding to the layer symbology (Section 8.4.3). For example, here is the HTML code that creates the first colored icon (dark red) in our legend:

```
<i style="background-color: #b30000"></i>
```

After the HTML is appended, the `<div>` element is returned with `return div;`. Lastly, the legend is added on the map using the `.addTo` method.

It is important to note that, in the above code, the legend is generated *manually*. In other words, the breaks and colors in the legend and in the map symobology (Section 8.4.3) are specified in two separate places in our code. It is up to us to make sure the labels and colors in the map legend indeed correspond to the ones we styled the layer with. A more general approach is to generate the legend *programmatically*, e.g., using a `for` loop going through the same **breaks** and **colors** arrays which we used when setting map symbology (Section 8.4.3 above). Here is an alternative version of the legend definition using a `for` loop, and the **breaks** and **colors** arrays[11] we defined in Section 8.4.3:

```
legend.onAdd = function(map) {
    var div = L.DomUtil.create("div", "legend");
    div.innerHTML =
         '<b>Population in 2015</b><br>by Town<br>' +
```

---

[11] Note that the loop goes over the **breaks** and **colors** arrays in reverse order, so that the legend entries are ordered from highest (on top) to lowest (at the bottom).

```
            '<small>Persons/Town</small><br>';
    for(var i = breaks.length-1; i > 0; i--) {
        div.innerHTML +=
            '<i style="background-color: ' + colors[i-1] + '"></i>' +
            breaks[i-1] + ' - ' + breaks[i] + '<br>';
    }
    return div;
};
```

With this alternative definition, the layer symbology and the legend are always in sync, since they are dependent on the same **breaks** and **colors** arrays. That way, changing the symbology (adding more breaks, changing the colors, etc.) will be automatically reflected both in town polygon colors and legend icon colors.

8.7 Using CSS to style the legend

One more thing we need to do regarding our legend is to give it the right placement and appearance, using CSS. The following CSS code is used to style our legend:

```
.legend {
    padding: 6px 8px;
    background-color: rgba(255,255,255,0.8);
    box-shadow: 0 0 15px rgba(0,0,0,0.2);
    border-radius: 5px;
}
.legend i {
    width: 18px;
    height: 18px;
    float: left;
    margin-right: 8px;
    opacity: 0.7;
}
div.legend.leaflet-control br {
    clear: both;
}
```

In the first CSS rule, we set properties for the legend as a whole, referring to `.legend`. We are setting padding, background color, box shadow, and border radius. In the second rule, we set our legend symbols (`<i>` elements) dimensions and also set `float: left;` so that the the symbols will be aligned into columns. Finally, the third rule makes sure the legend items are correctly aligned[12] regardless of browser zoom level.

The towns map `example-08-05.html`, now with a map legend, is shown in Figure 8.7.

---

[12]https://stackoverflow.com/questions/43793238/weird-leaflet-map-legend-display#answer-48345365

FIGURE 8.7: Screenshot of `example-08-05.html`

- Make a local copy of `example-08-05.html`.
- Replace the code section for defining the `getColor` function, to use a `for` loop instead of `if/else` conditionals. The alternative code section is given in Section 8.4.3. Don't forget to add the `breaks` and `colors` array definitions, in addition to the `for` loop!
- Replace the code section for generating HTML contents for the legend, to use a `for` loop instead of the fixed HTML string. The alternative code section is given in Section 8.6.
- You should see few small differences between the *labels* of each color category in the legends of `example-08-05.html` (Figure 8.7) and your modified version. Can you figure out which parts of the code are responsible for those differences?

8.8 Dynamic style

8.8.1 Styling in response to events

In the previous examples, we learned how to set a *constant* style, which is the same for all GeoJSON features (Section 8.3), and a *variable* style, which is dependent on feature properties, such as political party (Section 8.4.2) or population size (Section 8.4.3). In both cases, however, the style was determined on page load and remained the same, regardless of subsequent user interaction with the map. In this section, we will see how styling can also be made *dynamic*. This means the style can change while the user interacts with the page.

For example, a specific feature can be highlighted when the user places the mouse cursor above it (Figure 8.8). Dynamic styling can greatly enhance user experience and is one of the distinctive features of interactive web maps.

To achieve dynamic styling, we need to add event listeners to modify layer style in response to particular events. In the next example, we will add event listeners for changing any feature hovered with the mouse to "highlighted" style. Our event listeners are going to have to respond to `"mouseover"` and `"mouseout"` events (Section 4.8), and change the respective feature style to "highlighted" or "normal" state, respectively.

Specific event listeners per feature can be binded when loading the GeoJSON layer using the `onEachFeature` option, similarly to the way we used it to bind specific popups (Section 8.5). Inside the `onEachFeature` function, we can use the `.on` method of `layer` to add the event listeners. Our code for loading the GeoJSON layer then takes the following form, where `highlightFeature` and `resetHighlight`, as their name suggests, are functions for highlighting and resetting feature style:

```
$.getJSON("data/towns.geojson", function(data) {
    geojson = L.geoJSON(data, {
        style: style,
        onEachFeature: function(feature, layer) {
            layer
                .on("mouseover", highlightFeature)
                .on("mouseout", resetHighlight)
        }
    }).addTo(map);
});
```

The above code means that whenever we enter or leave a GeoJSON feature with the mouse, the `highlightFeature` or `resetHighlight` function will be executed, respectively. These functions (defined below) will be responsible for changing the feature style to "highlighted" or "normal", respectively.

How can we make sure we highlight the *specific* feature which triggered the event, i.e., the one we enter with the mouse, rather than any other feature? This brings us back to the **event object**, introduced in Section 4.11 and further discussed in Section 6.9. For example, in Section 6.9 we used the event object property named `.latlng` to obtain and display the clicked map coordinates within a popup (Section 6.15). In the present case, we use yet another property of the event object called `.target`, to get the reference to the page element which *triggered* the event, i.e., a reference to the hovered feature.

To understand what exactly the `.target` property contains when referring to a GeoJSON layer, we can manually create a reference to an individual feature in our GeoJSON layer `geojson` using the following expression, where 100 is the ID of a specific feature:

```
geojson._layers[100];
```

The ID values are arbitrary numbers, automatically generated by Leaflet. When necessary, however, we can always set our own IDs when the layer is created, again using the `onEachFeature` option (Section 8.8.3). The reference to a specific GeoJSON feature, such as `geojson._layers[100]`, contains numerous useful methods. Importantly, it has the following two methods that are useful for dynamic styling:

- `.setStyle`—Changes the style of the respective feature
- `.bringToFront`—Moves the feature to the front, in terms of display order

In addition, the entire GeoJSON layer object (such as `geojson`, in our example) has a method named `.resetStyle`. The `.resetStyle` method accepts a feature reference and resets the feature style back to the original GeoJSON style. Using these three methods, together with the `.target` property of the event object, our `highlightFeature` and `resetHighlight` functions can be defined as follows:

```
function highlightFeature(e) {
    var currentLayer = e.target;
    currentLayer.setStyle(highlightStyle);
    currentLayer.bringToFront();
}

function resetHighlight(e) {
    var currentLayer = e.target;
    geojson.resetStyle(currentLayer);
}
```

The `highlightStyle` object, which the `highlightFeature` function refers to, contains the "highlighted" style definition, separately defined below. The `highlightStyle` object specifies that the highlighted feature border becomes wider and its border color becomes yellow. The fill color of the highlighted feature also becomes more transparent, since in the default town polygon style we used an opacity of `0.7` (Section 8.4.3), whereas in the highlighted style we set fill opacity to `0.5`:

```
var highlightStyle = {
    weight: 5,
    color: "yellow",
    fillOpacity: 0.5
};
```

How do the `highlightFeature` and `resetHighlight` functions work in conjugation with the `highlightStyle` object? First of all, both functions accept the event object parameter `e`, which means the executed code can be customized to the particular properties of the event. Indeed, both functions use the `.target` property of the event object (`e.target`), which is a reference to the particular feature that triggered the event, i.e., the feature we are *entering* or *leaving* with the mouse cursor. That reference is captured in the `currentLayer` variable, for convenience. Determining the specific feature we wish to highlight or to "reset", hereby achieved with `e.target`, is crucial for dynamic styling.

On **mouse enter**, once the hovered feature was identified using `e.target`, the `highlightFeature` function uses the `.setStyle` method to set highlighted style (defined in `highlightStyle`) on the target feature. Then, the referenced feature is brought to front using the `.bringToFront` method, so that its borders—which are now highlighted—will not be obstructed by neighboring features on the map. On **mouse leave**, the `resetHighlight` function resets the styling using the `.resetStyle` method of the `geojson` layer, accepting the specific feature being "left" as its argument. This reverts the feature style back to the default one.

The resulting map `example-08-06.html` is shown in Figure 8.8. The town polygon being hovered with the mouse (e.g., Tel-Aviv) is highlighted in yellow.

FIGURE 8.8: Screenshot of `example-08-06.html`

8.8.2 Dynamic control contents

In addition to visually emphasizing the hovered feature, we may also want some other things to happen on our web page, reflecting the identity of the hovered feature in other ways. For example, we can have a dynamically updated **information box**, where relevant textual information about the feature is being displayed. This can be considered as an alternative to popups. The advantage of an information box over a popup is that the user does not need to *click* on each feature they want to get information on but just to *hover* with the mouse (Figure 8.9).

In the following `example-08-07.html`, we are going to add an information box displaying the name and population size of the currently hovered town, similarly to the information that was shown in the popups in `example-08-04.html` (Figure 8.6). The same technique used above to add a legend (Section 8.6) can be used to initialize the information box:

```
var info = L.control({position: "topright"});
info.onAdd = function(map) {
    var div = L.DomUtil.create("div", "info");
    div.innerHTML = '<h4>Towns in Israel</h4><p id="currentTown"></p>';
    return div;
};
info.addTo(map);
```

Initially, the information box contains just a heading ("Towns in Israel") and an *empty* paragraph with `id="currentTown"`. The paragraph will not always remain empty; it will be dynamically updated, using an event listener, every time the mouse cursor hovers over the towns layer. Before we go into the definition of the event listener, we will add some CSS code to make the information box look nicer, just like we did with the legend (Section 8.6):

```css
.info {
    padding: 6px 8px;
    font: 14px/16px Arial, Helvetica, sans-serif;
    background: rgba(255,255,255,0.8);
    box-shadow: 0 0 15px rgba(0,0,0,0.2);
    border-radius: 5px;
    width: 10em;
}
.info h4 {
    margin: 0 0 5px;
    color: #777;
}
.info #currentTown {
    margin: 6px 0;
}
```

As discussed in Section 8.8.1, the `.target` property of the event object—considering an event fired by a GeoJSON layer—is a reference to the specific feature being hovered with the mouse cursor. For instance, in `example-08-06.html` we used the `.setStyle` and the `.bringToFront` methods of the currently hovered feature to highlight it on mouse hover (Figure 8.8). The `.target` property of the event object also contains an internal property called `.feature`, which contains the specific GeoJSON *feature* (Section 7.3.3) the mouse pointer intersects, along with all of its properties. The expanded `highlightFeature` function (below) uses the **properties** of the GeoJSON feature to capture the name `name_eng` and population size value `pop_2015` of the currently hovered town. These two values are used to update the paragraph in the information box, using the `.html` method:

```javascript
function highlightFeature(e) {
    var currentLayer = e.target;
    currentLayer.setStyle(highlightStyle);
    currentLayer.bringToFront();
    $("#currentTown").html(
        currentLayer.feature.properties.name_eng + "<br>" +
        currentLayer.feature.properties.pop_2015 + " people"
    );
}
```

Accordingly, in the expanded version of the `resetHighlight` function, we now need to *clear* the text message when the mouse cursor leaves the feature. This can be done by setting the paragraph contents back to an empty string (`""`):

```javascript
function resetHighlight(e) {
    var currentLayer = e.target;
```

```
    geojson.resetStyle(currentLayer);
    $("#currentTown").html("");
}
```

The resulting map `example-08-07.html`, now with both the dynamic styling and the information box[13], is shown in Figure 8.9.

FIGURE 8.9: Screenshot of `example-08-07.html`

8.8.3   Linked views

There is an infinite amount of interactive behaviors that can be incorporated into web maps, the ones we covered in Sections 8.8.1–8.8.2 being just one example. The important take-home message from the example is that, using JavaScript and jQuery, the spatial entities displayed on our web map can be linked with other elements so that the map responds to user interaction in various ways. Moreover, the interaction is not necessarily limited to the map itself: we can associate the map with other elements on our page outside of the map `<div>`. This leads us to the idea of **linked views**.

Linked views are one of the most important concepts in interactive data visualization. The term refers to the situation when the same data are shown from different *points of view*, or different ways, while synchronizing user actions across all of those views. With web mapping, this usually means that in addition to a web map, our page contains one or more other panels, displaying information about the same spatial features in different ways: tables, graphs, lists, and so on. User selection on the map is reflected on the other panels, such as filtering the tables, highlighting data points on graphs, etc., and the other way around.

---

[13]Check out the *Leaflet Interactive Choropleth Map* tutorial (`https://leafletjs.com/examples/choropleth/`) for a walk-through of another example for setting symbology and interactive behavior in a Leaflet map.

The following `example-08-08.html` (Figure 8.10) implements a link between a web map and a *list*. Whenever a polygon of a given town is hovered on the map, the polygon itself as well as the corresponding entry in the towns list are highlighted. Similarly, whenever a town name is hovered on the list, the list item itself as well as the corresponding polygon are highlighted. This is mostly accomplished with methods we already covered in previous Chapters 4–7, except for several techniques[14]. The key to the association between the list and the GeoJSON layer is that both the list items and the GeoJSON features are assigned with matching IDs when loading the GeoJSON layer:

```javascript
function onEachFeature(feature, layer) {
    var town = feature.properties.town;
    var name_eng = feature.properties.name_eng;
    $("#townslist")
        .append('<li data-value="' + town + '">' + name_eng + '</li>');
    layer._leaflet_id = town;
    layer
        .on("mouseover", function(e) {
            var hovered_feature = e.target;
            hovered_feature.setStyle(highlightStyle);
            hovered_feature.bringToFront();
            $("li[data-value='" + hovered_feature._leaflet_id + "']")
                .addClass("highlight")[0]
                .scrollIntoView();
        })
        .on("mouseout", function(e) {
            var hovered_feature = e.target;
            geojson.resetStyle(hovered_feature);
            $("li[data-value='" + hovered_feature._leaflet_id + "']")
                .removeClass("highlight");
        });
}
```

In the above function, passed to the **onEachFeature** option and thus executed on each GeoJSON feature, the current town ID is captured in a variable called **town**. We use numeric town codes, which are stored in the **towns.geojson** file, in a GeoJSON property also named **town**:

```javascript
var town = feature.properties.town;
```

Then, the following expression assigns the town ID to each newly appended list item:

```javascript
$("#townslist")
    .append('<li data-value="' + town + '">' + name_eng + '</li>');
```

---

[14]This example is slightly more complex than other examples in the book, and we do not cover its code in as much detail as elsewhere in the book. It is provided mainly to demonstrate the idea of linked views and the principles of implementing it. Readers who are interested in using this approach in their work should carefully go over the code of `example-08-08.html` after reading this section.

The ID is stored in the `data-value` attribute of the `<li>` element. HTML attributes starting with `data-` are generic attributes for storing associated data[15] inside HTML elements. Correspondingly, the following expression assigns the same ID to the Leaflet layer feature:

```
layer._leaflet_id = town;
```

While the GeoJSON layer is being loaded, the following code section attaches event listeners for `"mouseover"` and `"mouseout"` on each feature:

```
layer
    .on("mouseover", function(e) {
        var hovered_feature = e.target;
        hovered_feature.setStyle(highlightStyle);
        hovered_feature.bringToFront();
        $("li[data-value='" + hovered_feature._leaflet_id + "']")
            .addClass("highlight")[0]
            .scrollIntoView();
    })
    .on("mouseout", function(e) {
        var hovered_feature = e.target;
        geojson.resetStyle(hovered_feature);
        $("li[data-value='" + hovered_feature._leaflet_id + "']")
            .removeClass("highlight");
    });
```

The event listeners take care of setting or resetting the polygon style, like we did in `example-08-06.html` (Section 8.8), and the `<li>` list item style, which is new to us. Note how the targeted `<li>` element corresponding to the hovered polygon is detected using the `hovered_feature._leaflet_id` property—this is the town ID which was assigned on GeoJSON layer load as shown above.

Once the layer is loaded, event listeners are binded to the list items too, with the following code section where `"#townslist li"` selector targets all `<li>` elements:

```
$("#townslist li")
    .on("mouseover", function(e) {
        var hovered_item = e.target;
        var hovered_id = $(hovered_item).data("value");
        $(hovered_item).addClass("highlight");
        geojson.getLayer(hovered_id).setStyle(highlightStyle);
    })
    .on("mouseout", function(e) {
        var hovered_item = e.target;
        var hovered_id = $(hovered_item).data("value");
        geojson.resetStyle(geojson.getLayer(hovered_id));
        $(hovered_item).removeClass("highlight");
    });
```

---

[15]https://developer.mozilla.org/en-US/docs/Learn/HTML/Howto/Use_data_attributes

FIGURE 8.10: Screenshot of `example-08-08.html`

These event listeners mirror the previous ones, meaning that now hovering on the list, rather than the layer, is the triggering event. This time, we determine the hovered ID using the `.data("value")` method, which returns the `data-value` attribute of the hovered `<li>` element.

The final result `example-08-08.html` is shown in Figure 8.10. All of the above may seem like a lot of work for a fairly simple effect, as we had to explicitly define each and every detail of the interactive behavior in our web map. However, consider the fact that using the presented methods there is complete freedom to build any type of innovative interactive behavior.

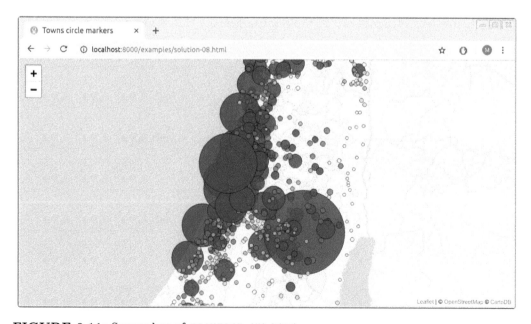

FIGURE 8.11: Screenshot of `solution-08.html`

8.9 Exercise

- Create a map of town population size, where circle markers are used instead of polygons, with circle *area* proportional to population size.
- To get a layer of town point locations, you can calculate the centroids of the `towns.geojson` layer in GIS software such as QGIS, or use the `towns_pnt.geojson` file provided with the online version of the book (Section 0.7).
- Circle area should be proportional to population size, therefore circle radius should be proportional to square root of population size. You can use the `Math.sqrt` function to calculate the square root of population size per town.
- In addition to `style` and `onEachFeature`, use the `pointToLayer` GeoJSON option as follows to display circle markers, instead of markers, so that the `radius` styling property is enabled: `pointToLayer: function(geoJsonPoint, latlng) { return L.circleMarker(latlng); }`.

Part III

Databases

9

Databases

9.1 Introduction

In Chapters 7–8, the foreground layers we displayed on Leaflet web maps came from GeoJSON files stored on the server. This is a viable approach when our data are relatively small and not constantly updated. When this is not the case, however, using GeoJSON files can become limiting.

For example, loading layers from GeoJSON files becomes prohibitive when files get too large, because the entire file needs to be transferred through the network, even if we only want to display just *some* of the content, for example by subsetting the layer in the JavaScript code after it has been received. It is obviously unreasonable to have the user wait until tens or hundreds of megabytes are being received, in the meanwhile seeing an empty map. Processing very large amounts of data can also make the browser unresponsive. A natural solution is to use a **database**. Unlike a file, a database can be queried to request just the minimal required portion of information each time, thus making sure that we are transferring and processing manageable amounts of data.

Another limitation of using GeoJSON files becomes apparent when the data are constantly updated and/or used for different purposes rather than just being displayed on a particular web map. For example, we may wish to build a web map displaying real-time municipal events, which means the data are constantly **updated** or edited (e.g., by the municipality staff) and/or used in various contexts (e.g., examined in GIS software by other professionals). Again, a natural solution is to use a database, shared between numerous concurrent connections for viewing and editing the data, through many types of different interfaces. Our web map, making use of one such concurrent connection, will therefore be synchronized with the database so that the displayed information is always up-to-date.

This chapter (Chapter 9) and the next two (Chapters 10–11) introduce the idea of loading data from a **spatial database** to display them on an interactive map, while dynamically filtering the data to transfer just the portion that we need. That way, we are freed from the limitation regarding the amount of data "behind" the web map. In other words, the database that stands behind our web map can be very large in size, yet the web map will stay responsive, thanks to the fact that we load subsets of the data each time, based on what the user chooses to see. In this chapter (Chapter 9), we introduce the concepts and technologies that enable a Leaflet map to load data from a spatial database. In the next two chapters, we go through examples of using non-spatial (Chapter 10) and spatial (Chapter 11) database queries for loading subsets of data from a database.

It should be mentioned that **Web Map Services (WMS)** (Section 6.6.5) comprise an alternative solution for displaying large, up-to-date amounts of data on a web map, however this solution is beyond the scope of this book. In short, with a WMS we are using a GIS database to build on-demand raster tiles. The server generates custom tiles based on the parameters it is given, so that the user has control of the displayed content, such as choosing which layers to display. This is unlike pre-compiled tile datasets, such as those introduced in Section 6.5.6 and used as base layers in the examples in Chapters 6–8, since pre-compiled tiles are fixed and cannot be dynamically modified based on user input.

There are valid use cases for both the database and WMS approaches. Basically, the database approach works better when loading vector layers that the user interacts with, which is made possible by the fact that the server can send raw data (such as GeoJSON), and we can control the way that data are displayed on the client, using JavaScript code. The WMS approach works better when our data are very complex and have elaborate symbology. In such cases, it makes sense to have a dedicated map server with specialized software to build raster images with the displayed content, and send them to the client to be displayed as-is[1].

9.2   What is CARTO?

9.2.1   The CARTO platform

A problem that immediately arises regarding retrieval of spatial data from a database onto a web map is that client-side scripts cannot directly connect to a database. A dynamic server, which we mentioned in Section 5.4.3, is the solution to this problem. On the dynamic server, server-side scripts, which indeed *can* connect to the database, are used to query the database and send the data back to the client. In fact, the need to send information from a database to the browser is one of the main motives for setting up a dynamic server.

In this book, we focus on client-side solutions, so we will not be dealing with setting up our own dynamic server coupled with a database. Instead, we will use a cloud-based service by a company called **CARTO**[2]. The CARTO platform provides several cloud computing GIS and web-mapping services. One of the most notable services, and the one we are going to use in this book, is the **SQL API** (Section 9.7). CARTO allows you to upload your own data into a managed spatial database, while CARTO's SQL API allows you to interact with that database. In other words, CARTO takes care of setting up and maintaining a spatial database, as well as setting up server-side components to make that database reachable through HTTP.

In this chapter, we will introduce the CARTO platform and the technologies it is based on: databases (Section 9.3) and spatial databases (Section 9.4), PostGIS (Section 9.5), SQL (Section 9.6), and the SQL API (Section 9.7). Towards the end of the chapter, we will see how CARTO can be used for querying and displaying data from a database on a Leaflet map (Section 9.8). In the next two chapters, we will dig a little deeper into different types of queries and their utilization in web mapping. In Chapter 10, we will see an example of non-spatial, attribute-based filtering of data, based on user input from a dropdown menu.

---

[1] There is an official tutorial on using *WMS with Leaflet* (`https://leafletjs.com/examples/wms/wms.html`), where you can see a practical example.
[2] `https://carto.com/`

In Chapter 11, we will see an example of using spatial queries to retrieve data based on proximity to a clicked location.

9.2.2 Alternatives to CARTO

CARTO is a commercial service that comes at a price[3], though at the time of writing there is an option called **Student Developer Pack**[4] which includes a two-year free CARTO account. It is worth mentioning that the CARTO platform is **open-source**[5]. In principle, it can be installed on any computer to replicate almost[6] the entire functionality of CARTO for free. The installation and maintenance are quite complicated though.

A minimal alternative to CARTO, comprising a simple dynamic server, a database, and an SQL API can be set up (relatively) more easily. The online version of the book (Section 0.7) includes an additional supplement with instructions for one way to do that, using **Node.js** and **PostgreSQL/PostGIS**. This is beyond the scope of the main text of the book, but it is important to be aware that no matter how convenient CARTO is, if some day we need to cut costs and manage our own server for the tasks covered in the following chapters—it can be done. Accordingly, all examples where a CARTO database is used are also given in alternative versions, using the custom SQL API setup instead of CARTO. The latter are marked with `-s`, as in `example-09-01.html` and `example-09-01-s.html` (Appendix B).

9.3 Databases

The term **database**[7] describes an organized collection of data. A database stores data, but also facilitates indexing, searching, and querying the data, as well as modifying and adding new data. The most established and commonly used databases follow the **relational model**[8], where the records are organized in tables, and the tables are usually associated with one another via common columns.

For example, Figure 9.1 shows a small hypothetical database with two tables named `airports` and `flights`. The airports table gives the `name` and location (`lon`, `lat`) of seven different airports. The `flights` table lists the departure time (`year`, `month`, `day`, `dep_time`), the origin (`origin`), and the destination (`dest`) of five different flights that took place between the airports listed in the `airports` table on a particular day. Importantly, the `airports` and `flights` tables are related through the *airport code* column `faa` in the `airports` table matching the `origin` and `dest` columns in the `flights` table.

Relational database queries, including both "ordinary" (Section 9.6.2) and spatial (Sections 9.6.3–9.6.4) queries, are expressed in a language called SQL (Section 9.6).

---

[3]https://carto.com/pricing/
[4]https://education.github.com/pack
[5]https://github.com/CartoDB/cartodb
[6]https://gis.stackexchange.com/questions/269822/cartodb-opensource-vs-paid-one
[7]https://en.wikipedia.org/wiki/Database
[8]https://en.wikipedia.org/wiki/Relational_database

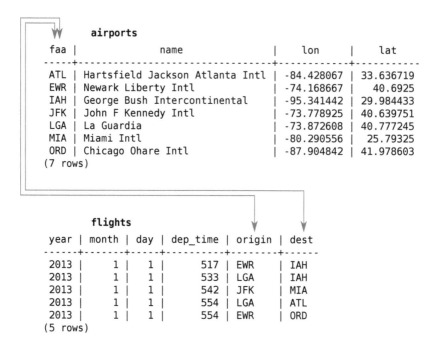

FIGURE 9.1: An example of a relational database with two tables

9.4   Spatial databases

A **spatial database**[9] is a database that is optimized to store and query data that represents objects defined in a geometric space. Regarding the *storage* part, plainly speaking, the tables in a spatial database have a special type of **geometry column**, which holds the geometric component of that specific record, i.e., the geometry type and the coordinates. This may sound familiar—recall that geometry GeoJSON types (Section 7.3.2), which represent just the geometric part of a feature, consist of two properties: `"type"` and `"coordinates"`. The similarity between the geometry column and the GeoJSON geometry types in not incidental, but due to the fact that both are based on the Simple Features standard, which we mentioned in Section 7.3.1. The difference is that in a spatial database, the geometries are usually encoded in a format called **Well-Known Binary (WKB)**, a binary version of the Well-Known Text (WKT) format (which we mentioned in Section 7.3.1), rather than in the GeoJSON format.

In addition to geometry storage, spatial databases define special functions that allow for queries based on *geometry*. This means we can use the database to make spatial numeric calculations (e.g., geographical distance; Sections 11.3.2–11.3.3), retrieve data based on location (e.g., K-nearest neighbors; Section 11.3.2), or create new geometries (e.g., calculating the centroid of a geometry). These are called **spatial queries** (Section 9.6.4), since they involve the spatial component of the database, i.e., the geometry column of at least one

---

[9]https://en.wikipedia.org/wiki/Spatial_database

table. The concept is very similar to spatial operators and functions used in GIS software, such as the *Select by Location*[10] tool in ArcGIS.

Commonly used open-source spatial databases include **PostgreSQL/PostGIS** (used by CARTO; see Section 9.5 below), **SQLite/SpatiaLite**[11], and **MySQL**[12]. There are also proprietary databases that support spatial data, such as **Oracle Spatial**[13] and **Microsoft SQL server**[14].

9.5 What is PostGIS?

PostGIS[15] is a popular extension for the **PostgreSQL**[16] database, making the PostgreSQL/PostGIS combination a spatial database (Obe and Hsu, 2015). In other words, a PostgreSQL database with the PostGIS extension enabled allows for storage of spatial data and execution of spatial SQL queries (Section 9.6.4). At the moment, the PostgreSQL/PostGIS combination[17] makes the most powerful open-source spatial database available.

Since both PostgreSQL and PostGIS are free and open-source, you can install PostGIS on your computer and set up your own database. However, running a database requires some advanced setup and maintenance, which is outside the scope of this book[18]. As mentioned previously, the CARTO platform provides a cloud-based PostGIS database as a service, which we are going to use in this book.

9.6 What is SQL?

9.6.1 Overview

SQL[19] is a language for writing statements to query or to modify tables stored in a relational database, whether spatial or non-spatial. Using SQL, you can perform many types of tasks: filtering, joining, inserting new data, updating existing data, etc. SQL statements can be executed in many types of database **interfaces**, from command lines interfaces, through database administrator consoles in GIS software, and to APIs that connect to the database

---

[10]http://desktop.arcgis.com/en/arcmap/10.3/map/working-with-layers/using-select-by-location.htm
[11]https://en.wikipedia.org/wiki/SpatiaLite
[12]https://dev.mysql.com/doc/refman/8.0/en/spatial-types.html
[13]https://en.wikipedia.org/wiki/Oracle_Spatial_and_Graph
[14]https://docs.microsoft.com/en-us/sql/relational-databases/spatial/spatial-data-sql-server
[15]https://en.wikipedia.org/wiki/PostGIS
[16]https://en.wikipedia.org/wiki/PostgreSQL
[17]PostgreSQL with the PostGIS extension will be referred to as *PostGIS* from now on, for simplicity.
[18]The online supplement includes instructions for installing PostGIS, as part of the custom SQL API setup (Section 9.2.2).
[19]https://en.wikipedia.org/wiki/SQL

through HTTP—such as the CARTO SQL API, which we are going to use (Section 9.7). You may already be familiar with SQL syntax from GIS software, such as **ArcGIS**[20] and **QGIS**[21], where SQL can be used to select features from a spatial layer.

Using CARTO, we will experiment with writing SQL queries to extract data from a cloud-based database (Section 9.7) and to display these data on Leaflet map (Section 9.8). That way, we will become familiar with the whole idea of querying spatial databases, from the web-mapping perspective. Hopefully, this introduction will be of use if, later on, you decide to go deeper into the subject and set up a spatial database on your own (Section 9.2.2).

SQL, as you can imagine, is a very large topic (Nield, 2016; DeBarros, 2018). The syntax of SQL is not the focus of this book, so we will not go deeply into details nor will we cover the whole range of query types that can be used for various tasks. In the following Chapters 10–11, we will only encounter about ~5-6 relatively simple types of SQL queries, most of which are briefly introduced below (Sections 9.6.2–9.6.4). This set of SQL queries will be enough for our purposes, and you will be able to modify the code to apply the same type of queries to different data, even if you have never used SQL before.

In the following examples (Sections 9.6.2–9.6.4), we will demonstrate several types of SQL queries on a database that contains just one table, a table named `plants`. The `plants` table contains rare plant observations in Israel[22]. Each row in the `plants` table represents an individual observation of a rare plant species. The table has different columns describing each observation, such as:

- `id`—Observation ID
- `name_lat`—Latin species name
- `obsr_date`—Observation date
- `geometry`—The location; this column is the **geometry column** (Section 9.4)

The query examples are just for illustration and are not meant to be replicated in a console or command line, since we are not setting up our own database. However, shortly you will be able to execute them through the CARTO platform (Section 9.7).

9.6.2   Non-spatial queries

The most basic SQL statement is the `SELECT` statement. A `SELECT` statement pulls data from a table, possibly filtered on various criteria and supplemented with new columns resulting from table joins or transformations. For example, we can use the following `SELECT` query to get a subset of the `plants` table, with just three of its columns: `id`, `name_lat` and `obsr_date`. The table is also filtered, to include only those rows where the Latin species name is equal to the specific value `'Anticharis glandulosa'`:

```
SELECT id, name_lat, obsr_date
  FROM plants
  WHERE name_lat = 'Anticharis glandulosa';
```

---

[20]http://pro.arcgis.com/en/pro-app/help/mapping/navigation/sql-reference-for-elements-used-in-query-expressions.htm

[21]https://docs.qgis.org/2.8/en/docs/user_manual/working_with_vector/query_builder.html

[22]The data source of this table is the *Endangered Plants of Israel* (https://redlist.parks.org.il/) website by the Israel Nature and Parks Authority.

By convention, SQL keywords are written in uppercase, while specific values—such as column names—are written in lowercase. This is not strictly required, as SQL is not case-sensitive, unlike JavaScript, for instance, which *is* case-sensitive. Spaces and line breaks are ignored in SQL, like in JavaScript. The query ends with the ; symbol.

Note the way that the query is structured. The queried column names are listed after the `SELECT` keyword, the table name is specified after `FROM`, and the condition for filtering returned records is constructed after the `WHERE` keyword. In this example, the condition `name_lat='Anticharis glandulosa'` means "return all records (rows) where the value of `name_lat` is equal to `'Anticharis glandulosa'`".

If we had access to a PostGIS database with the `plants` table and could type the above SQL query through its command line interface (called **psql**[23]), the following textual printout with the query result would appear in the command line[24]:

```
   id   |        name_lat        |  obsr_date
--------+------------------------+------------
 339632 | Anticharis glandulosa  | 1988-03-18
 359135 | Anticharis glandulosa  | 2012-12-15
 367327 | Anticharis glandulosa  | 2012-12-15
(3 rows)
```

According to the result, we can tell that there are only three observations of `'Anticharis glandulosa'` in the `plants` table. Note that the last line is not part of the result, but only specifies the number of returned rows.

9.6.3 The geometry column

As mentioned in Section 9.4, the distinctive feature of a spatial database is that its tables may contain a geometry column. The values in the geometry column specify the spatial locations of the respective database records (i.e., the table rows). The geometry column usually contains binary code, which is an encoded version of the Well-Known Text (WKT) format, known as Well-Known Binary (WKB). The binary compression is conventionally used to reduce the required storage space for the database.

For example, the geometry column in our `plants` table is named `geometry`. The following query returns the contents of three columns from the `plants` table, the "ordinary" id and `name_lat` columns, as well as the geometry column named `geometry`. The query is also limited to the first five records, with the `LIMIT 5` part:

```
SELECT id, name_lat, geometry
  FROM plants
  LIMIT 5;
```

Here is the printout we would see on the command line in this case:

```
   id   |   name_lat    |                   geometry
--------+---------------+-----------------------------------------------------
 321432 | Iris haynei   | 0101000000520C906802D741400249D8B793624040
```

---

[23]http://postgresguide.com/utilities/psql.html
[24]Instructions to set up a PostGIS database and import the `plants` table are given in the online supplement, as part of setting up an alternative SQL API (Section 9.2.2).

```
321433 | Iris haynei    | 0101000000D235936FB6D34140C6151747E55E4040
321456 | Iris atrofusca | 01010000001590F63FC0984140EDB60BCD75723F40
321457 | Iris atrofusca | 0101000000672618CE35984140357C0BEBC6833F40
321459 | Iris vartanii  | 0101000000E6B0FB8EE19141405D6E30D461793F40
(5 rows)
```

It is evident the WKB strings in the `geometry` column make no sense to the human eye. However, WKB can always be converted to its textual counterpart **WKT**, using the `ST_AsText` operator, as demonstrated in the following, slightly modified, version of the above SQL query:

```
SELECT id, name_lat, ST_AsText(geometry) AS geom
  FROM plants
  LIMIT 5;
```

In the modified query, we replaced the `geometry` part with `ST_AsText(geometry)`, thus *transforming* the column from WKB to WKT. The `AS geom` part is used to rename the new column to `geom` (otherwise it would get a default name such as `st_astext`). Here is the resulting table, with the geometry column transformed to its WKT representation and renamed to `geom`:

```
  id   |   name_lat     |             geom
--------+----------------+----------------------------
 321432 | Iris haynei    | POINT(35.679761 32.770133)
 321433 | Iris haynei    | POINT(35.654005 32.741372)
 321456 | Iris atrofusca | POINT(35.193367 31.44711)
 321457 | Iris atrofusca | POINT(35.189142 31.514754)
 321459 | Iris vartanii  | POINT(35.139696 31.474149)
(5 rows)
```

Similarly, we can convert the WKB geometry column to **GeoJSON**, which we are familiar with from Chapter 7. To do that, we simply use the `ST_AsGeoJSON` function instead of the `ST_AsText` function, as follows:

```
SELECT id, name_lat, ST_AsGeoJSON(geometry) AS geom
  FROM plants
  LIMIT 5;
```

Here is the result, with the geometry column now given in the GeoJSON format:

```
  id   |   name_lat     |                          geom
--------+----------------+--------------------------------------------------------------
 321432 | Iris haynei    | {"type":"Point","coordinates":[35.679761,32.770133]}
 321433 | Iris haynei    | {"type":"Point","coordinates":[35.654005,32.741372]}
 321456 | Iris atrofusca | {"type":"Point","coordinates":[35.193367,31.44711]}
 321457 | Iris atrofusca | {"type":"Point","coordinates":[35.189142,31.514754]}
 321459 | Iris vartanii  | {"type":"Point","coordinates":[35.139696,31.474149]}
(5 rows)
```

Examining either one of the last two query results, we can tell that the the `plants` table—or at least its first five records—contains geometries of type "Point" (Table 7.2).

9.6.4 Spatial queries

The geometry column can be used to apply **spatial operators** on our table, just like in GIS software. Much like general SQL (shown previously), the syntax of spatial SQL queries is a very large topic (Obe and Hsu, 2015), and mostly beyond the scope of this book. In Chapter 11 we will experiment with just one type of a spatial query, which returns the *nearest* records from a given point.

For example, the following spatial query returns the nearest five observations from the `plants` table based on distance to the specific point `[34.810696, 31.895923]` (as in `[lon, lat]`). Plainly speaking, this SQL query *sorts* the entire `plants` table by decreasing proximity of all geometries to `[34.810696, 31.895923]`, then the top five records are returned:

```
SELECT id, name_lat, ST_AsText(geometry) AS geom
  FROM plants
  ORDER BY
    geometry::geography <->
    ST_SetSRID(
      ST_MakePoint(34.810696, 31.895923), 4326
    )::geography
  LIMIT 5;
```

The selection of top five records is done using the `LIMIT 5` part. The spatial operators part comes after the `ORDER BY` keyword, where we calculate all distances from `plants` points to a specific point `[34.810696, 31.895923]`, and use those distances to sort the table. We will elaborate on this part in Chapter 11. Here is the result, with the five nearest observations to `[34.810696, 31.895923]`:

```
   id    |       name_lat       |          geom
---------+----------------------+--------------------------
 341210  | Lavandula stoechas   | POINT(34.808564 31.897377)
 368026  | Bunium ferulaceum    | POINT(34.808504 31.897328)
 332743  | Bunium ferulaceum    | POINT(34.808504 31.897328)
 328390  | Silene modesta       | POINT(34.822295 31.900125)
 360546  | Corrigiola litoralis | POINT(34.825931 31.900792)
(5 rows)
```

For more information, Chapter 7 in the *Introduction to Data Technologies* book (Murrell, 2009) gives a good introduction to (non-spatial) SQL. The *W3Schools SQL Tutorial*[25] can also be useful for quick reference of commonly used SQL commands. An introduction to spatial operators and PostGIS can be found in the official *Introduction to PostGIS*[26] tutorial and in the *PostGIS in Action* book (Obe and Hsu, 2015).

---

[25]https://www.w3schools.com/sql/
[26]https://postgis.net/workshops/postgis-intro/

9.7   The CARTO SQL API

9.7.1   API usage

The **CARTO SQL API**[27] is an API for communication between a program that understands HTTP, such as the browser, and a PostGIS database hosted on the CARTO platform. The CARTO SQL API allows for users to send SQL queries to their PostGIS database on the CARTO platform. The queries are sent via HTTP (Section 5.3), typically by making a `GET` request (Section 5.3.2.2) using a URL which includes the CARTO user name and the SQL query. The CARTO server processes the request and prepares the returned data, according to the SQL query applied on the particular user's database. The result is then sent back, in a format of choice, such as **CSV**, **JSON**, or **GeoJSON**. Importantly, the fact that the requests are made through HTTP means that we can send requests to the database, and get the responses, from client-side JavaScript code using Ajax (Section 7.7).

It is important to note that some types of queries other than `SELECT`, namely queries that modify our data, such as `INSERT`, require an **API key** as an additional parameter in our request. The API key serves as a password for making sensitive queries. Without this restriction, anyone who knows our username could make changes to our database, or even delete its entire contents. We will not be using this type of sensitive queries up until Section 13.6.1, where a method to hide the API key from the client and still be able to write to the database will be introduced.

The basic URL structure for sending a `GET` request to the CARTO SQL API looks like this:

`https://CARTO_USERNAME.carto.com/api/v2/sql?format=FORMAT&q=SQL_STATEMENT`

where:

- `CARTO_USERNAME` should be replaced with your CARTO **user name**
- `FORMAT` should be replaced with the required **format**
- `SQL_STATEMENT` should be replaced with the SQL **query**

Note that this is a special URL structure, which contains a **query string**[28]. A query string is used to send parameters to the server as part of the URL. The query string comes at the end of the URL, after the ? symbol, with the parameters separated by & symbols. In this case, the query string contains two parameters, `format` and `q`.

For example, here is a specific query:

`https://michaeldorman.carto.com/api/v2/sql?format=GeoJSON&q=`
`SELECT id, name_lat, the_geom FROM plants LIMIT 2`

where:

- `CARTO_USERNAME` was replaced with `michaeldorman`
- `FORMAT` was replaced with `GeoJSON`
- `SQL_STATEMENT` was replaced with `SELECT id, name_lat, the_geom FROM plants LIMIT 2`

---

[27]https://carto.com/developers/sql-api/
[28]https://en.wikipedia.org/wiki/Query_string

Possible values for the `format` parameter include `JSON`, `GeoJSON`, and `CSV`. The default returned format is `JSON`, so to get your result returned in JSON you can omit the `format` parameter to get a slightly simplified URL:

```
https://michaeldorman.carto.com/api/v2/sql?q=
SELECT id, name_lat, the_geom FROM plants LIMIT 2
```

To get your results in a format other than JSON, such as GeoJSON or CSV, you need to explicitly specify the format, as in `format=GeoJSON` or `format=CSV`. Other possible formats include `GPKG`, `SHP`, `SVG`, `KML`, and `SpatiaLite`[29].

9.7.2 Query example

The `plants` table used in the above SQL examples (Section 9.6) was already uploaded to a CARTO account named `michaeldorman`. We will query the database associated with this account to experiment with the CARTO SQL API.

Figure 9.2 shows how the `plants` table appears on the CARTO web interface. Note the table name—`plants`, in the top-left corner—and column names—such as `id` and `name_lat`—which we referred to when constructing the SQL queries (Section 9.6). Importantly, note the geometry column—the one with the small **GEO** icon next to it—named `the_geom`[30]. We will shortly go over the procedure of uploading data to your own CARTO account (Section 9.7.3).

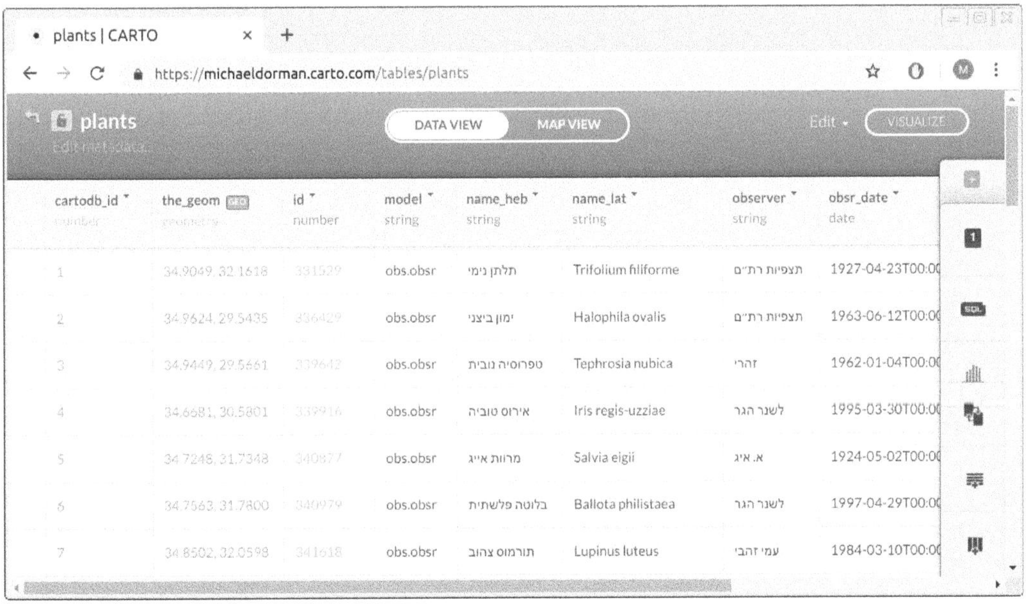

FIGURE 9.2: The `plants` table, as displayed in the CARTO web interface

---

[29]For the full list, refer to the *CARTO SQL API documentation* (`https://carto.com/developers/sql-api/reference/`).

[30]The CARTO platform conventionally uses the name `the_geom` for the geometry column. In principle, the geometry column can be named any other way. For example, in the SQL examples in Section 9.6 the geometry column was named `geometry`, which is another commonly used convention.

Let's try to send a query to the CARTO SQL API to get some data, in the GeoJSON format, from the **plants** table. Paste the following query into the browser's address bar:

```
https://michaeldorman.carto.com/api/v2/sql?format=GeoJSON&q=
SELECT id, name_lat, the_geom FROM plants LIMIT 2
```

Since we specified **format=GeoJSON**, a GeoJSON file will be returned (Section 9.7.1). The query q was **SELECT id, name_lat, the_geom FROM plants LIMIT 2**, which means that we request the **id**, **name_lat** and **the_geom** columns from the **plants** table, limited to the first 2 records (Section 9.6.3). As a result, the CARTO server takes the relevant information from the **plants** table and returns the following GeoJSON content:

```
{
  "type": "FeatureCollection",
  "features": [
    {
      "type": "Feature",
      "geometry": {
        "type": "Point",
        "coordinates": [35.032018, 32.800539]
      },
      "properties": {
        "id": 345305,
        "name_lat": "Elymus elongatus"
      }
    },
    {
      "type": "Feature",
      "geometry": {
        "type": "Point",
        "coordinates": [35.564703, 33.047197]
      },
      "properties": {
        "id": 346805,
        "name_lat": "Galium chaetopodum"
      }
    }
  ]
}
```

This is a GeoJSON string of type **"FeatureCollection"** (Section 7.3.4). It contains two features with **"Point"** geometries, each having two non-spatial attributes: **id** and **name_lat**.

Remember that the geometry column **the_geom** needs to appear in the query whenever we export the result in a spatial format, such as **format=GeoJSON**. Otherwise, the server cannot generate the geometric part of the layer and we get an error. For example, omitting the **the_geom** column from the above query:

```
https://michaeldorman.carto.com/api/v2/sql?format=GeoJSON&q=
SELECT id, name_lat FROM plants LIMIT 2
```

returns the following error message instead of the requested GeoJSON:

```
{"error":["column \"the_geom\" does not exist"]}
```

By the way, while pasting these URL examples into the browser, you may have noticed how the browser automatically **encodes**[31] the URL into a format that can be transmitted over the internet. This is something that happens automatically, and we do not need to worry about. For example, as part of URL encoding, spaces are converted to %20, so that the URL we typed above:

```
https://michaeldorman.carto.com/api/v2/sql?format=GeoJSON&q=
SELECT id, name_lat, the_geom FROM plants LIMIT 2
```

becomes:

```
https://michaeldorman.carto.com/api/v2/sql?format=GeoJSON&q=
SELECT%20id,%20name_lat,%20the_geom%20FROM%20plants%20LIMIT%202
```

Since the returned file is in the GeoJSON format, we can immediately import it into various spatial applications. For example, the file can be displayed and inspected in GIS software such as **QGIS** (Figure 9.3). If you are not using GIS software, you can still examine the GeoJSON file by importing it into the **geojson.io** web interface (Section 7.4.1). More importantly for our cause, the GeoJSON content can be instantly loaded in a Leaflet web map, as will be demonstrated next in Section 9.8.

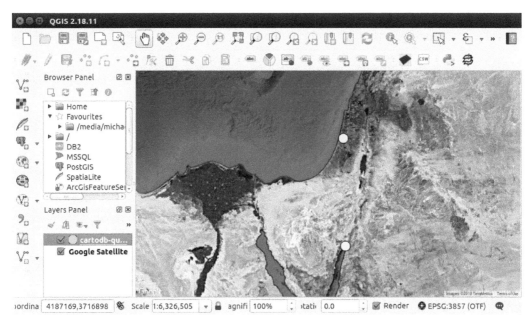

FIGURE 9.3: A GeoJSON file, obtained from the CARTO SQL API, displayed in QGIS

Exporting data in the JSON format is very similar to GeoJSON, but applicable for *non-spatial* queries that cannot be converted to GeoJSON. We will see a practical example of exporting JSON data from CARTO in Section 10.4.3.

- Try executing the non-spatial SQL query example from Section 9.6.2 with the CARTO SQL API, using **JSON** as the export format.

---

[31]https://en.wikipedia.org/wiki/Percent-encoding

Lastly, as an example of the **CSV** export format, try the following request to the SQL API:

```
https://michaeldorman.carto.com/api/v2/sql?format=CSV&q=
SELECT id, name_lat, obsr_date, ST_AsText(the_geom) AS geom
FROM plants WHERE name_lat = 'Iris mariae' LIMIT 3
```

Note that we use the `format=CSV` parameter so that the result comes as a CSV file. CSV is a plain-text tabular format. Given the above query, the resulting CSV file contains the following text:

```
id,      name_lat,     obsr_date,  geom
358449, Iris mariae, 2010-03-13, POINT(34.412502 31.019879)
359806, Iris mariae, 2015-03-08, POINT(34.713009 30.972615)
337260, Iris mariae, 2001-02-23, POINT(34.63678 30.92807)
```

The CSV file can also be opened in a spreadsheet software such as **Microsoft Excel** or **LibreOffice Calc** (Figure 9.4)[32].

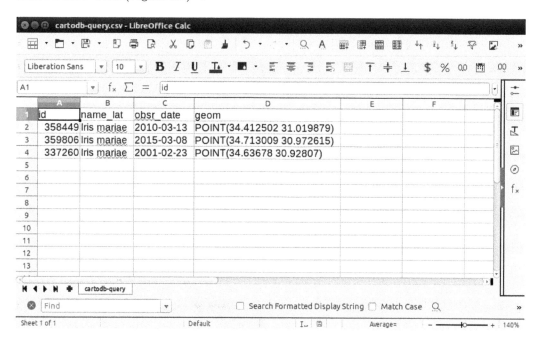

FIGURE 9.4: CSV file exported from the CARTO SQL API displayed in a spreadsheet software (LibreOffice Calc)

9.7.3 Uploading your data

Before we begin with connecting a Leaflet map with data from CARTO (Section 9.8), you may want to experiment with your own account, possibly with different data instead of the `plants` table. Assuming you already signed up on `https://carto.com/` and have a CARTO account, the easiest way to upload data is the use the CARTO web interface, as follows:

- Go to `https://carto.com/` and log in to your account.

---

[32]For a more detailed description of the CARTO SQL API, see the *documentation* (`https://carto.com/developers/sql-api/`).

- Once you are in your user's home page, click on **Data** in the upper-left corner. This will show the different tables in your database on CARTO. For example, Figure 9.5 shows the datasets page with three tables, named `beer_sheva`, `plants`, and `earthquake_sql`. This screen may be empty if you have just created a new CARTO account and have not uploaded any data yet.
- Click on the **New Dataset** button in the top-right corner of the screen.
- You will see different buttons for various methods of importing data. The simplest option is to upload a GeoJSON file. Choose the **Data file** option in the upper ribbon, then click on the **BROWSE** button and navigate to your GeoJSON file. Finally, click on the **CONNECT DATASET** button (Figure 9.6). You can upload the `plants.geojson` file from the book materials (Section A) into your own account to experiment with the same dataset as shown in the examples.
- Once the file is uploaded, you should be able to see it as a new table in your list of datasets on CARTO. You can view the table in the CARTO web interface (Figure 9.2), and even edit its contents. For example, you can change the table name, rename any of the columns, edit cell contents, add new rows, etc.

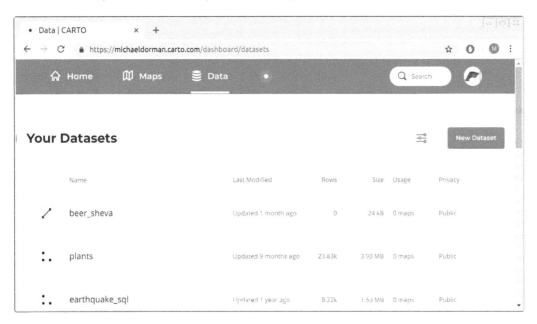

FIGURE 9.5: Datasets screen on CARTO

- Upload any GeoJSON file other than `plants.gejson` to CARTO, then try to adapt the above SQL API queries to your own username, table name, and column names.
- Test the new queries by pasting the respective URLs into the browser address bar and examining the returned results.

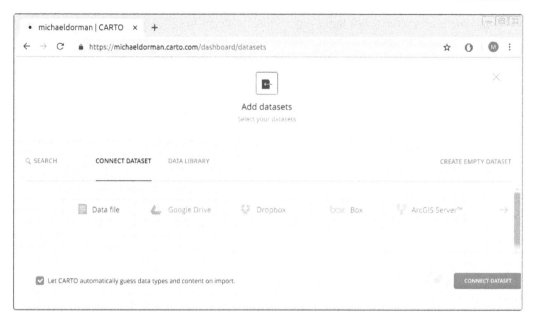

FIGURE 9.6: The file upload screen in the CARTO web interface

9.8   CARTO and Leaflet

We have just learned how to use the CARTO SQL API to send SQL queries to a CARTO database (Section 9.7). Importantly, since we are working with a *spatial* database, one of the formats in which we can choose to get the returned results is GeoJSON (Section 9.7.2). In this section, we will load a CARTO query result in a web page and display it on a Leaflet map. The method we are going to use for loading the query result is the `$.getJSON` function, which we introduced in Section 7.8 and used in many of the examples in Chapters 7–8, for loading GeoJSON layers from files.

Our starting point is the basic map `example-06-02.html` from Section 6.5.7, with two small preliminary changes. First, we include the jQuery library in the `<head>`, since we will use the `$.getJSON` function from that library:

```
<script src="js/jquery.js"></script>
```

Second, we change the initial map extent as follows, so that the `plants` observations will be visible on page load:

```
var map = L.map("map").setView([32, 35], 8);
```

Now, in order to load data from CARTO on the Leaflet map, we need to go through the following steps:

- Construct the URL to query the CARTO SQL API
- Get the data from CARTO and add it on the map

As the first step, we will construct the query URL. For convenience, the URL will be split in two parts: the fixed **base URL** prefix and the varying **SQL query** suffix. Combining both parts gives the complete URL, which we will use to retrieve data from the CARTO database. The fixed base URL, specific to a particular CARTO user, can be kept in a separate variable hereby named `url`. That way, we do not need to repeat it in each and every query we make in our script:

```
var url = "https://michaeldorman.carto.com/api/v2/sql?format=GeoJSON&q=";
```

Note that to make the code even more manageable you can split the base URL into two parts too, keeping the user name in a separate variable. That way, if we need to switch to a different CARTO account it is more clear which part of the code needs to be modified:

```
var cartoUserName = "michaeldorman";
var url =
    "https://" + cartoUserName + ".carto.com/api/v2/sql?format=GeoJSON&q=";
```

Either way, our next step is to define the varying SQL query part, used to retrieve data from the database according to a specific query. For example, we can use the following query, which returns the `name_lat` and `the_geom` columns for the first 25 records from the `plants` table:

```
var sqlQuery = "SELECT name_lat, the_geom FROM plants LIMIT 25";
```

Remember that you need to include the geometry column in your query whenever the requested format is GeoJSON. Otherwise, the layer cannot be generated and we get an error (Section 9.7.2). When the base URL and the SQL query are combined, using `url+sqlQuery`, we get the complete URL:

```
https://michaeldorman.carto.com/api/v2/sql?format=GeoJSON&q=
SELECT name_lat, the_geom FROM plants LIMIT 25
```

The complete URL can be passed to `$.getJSON` to load the resulting GeoJSON from CARTO on the Leaflet map:

```
$.getJSON(url + sqlQuery, function(data) {
    L.geoJSON(data, {
        onEachFeature: function(feature, layer) {
            layer.bindPopup(feature.properties.name_lat);
        }
    }).addTo(map);
});
```

This code should be familiar from Chapters 7–8. The outermost function is `$.getJSON`, which we use to make an Ajax `GET` request from another location on the internet (CARTO). Since the returned data are in the GeoJSON format (as specified with `format=GeoJSON`), the callback function of `$.getJSON` can use the `L.geoJSON` function to immediately convert the GeoJSON object to a Leaflet GeoJSON layer. Using the `onEachFeature` option we are also binding specific popups (Section 8.5) for each feature to display the Latin name of the observed plant species. Finally, the layer is added on the map with the `.addTo` method.

The resulting map `example-09-01.html` is shown in Figure 9.7. Our data from CARTO, i.e., the first 25 plant observations, are loaded on the map!

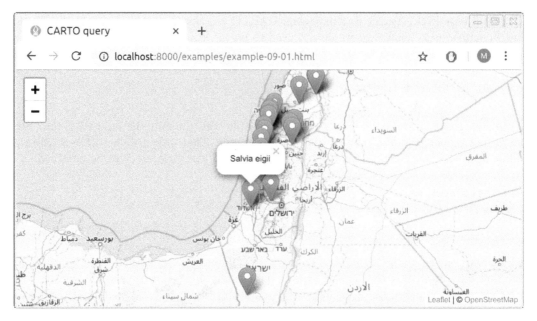

FIGURE 9.7: Screenshot of `example-09-01.html`

- Paste the above code section into the console of `example-09-01.html`.
- Modify the SQL query (`sqlQuery`) to experiment with adding different observations on the map, according to the SQL examples shown in Section 9.6.
- For example, you can replace the `LIMIT 25` part with a condition of the form `WHERE name_lat = '...'` to load all observations of a particular species (Section 9.6.2).

We have now covered the general principles of using the CARTO SQL API to display layers coming from a database on a Leaflet map. So far, however, what we did was not very different from loading a GeoJSON *file* on a map, like we did in Chapters 7–8. The only difference is that the path to the GeoJSON file was a URL addressing the CARTO SQL API, rather than a local (Section 7.8.2) or remote (Section 7.8.3) GeoJSON file. Still, the query was fixed, in the sense that exactly the same layer with 25 observations (Figure 9.7) will be displayed each time the page is loaded (unless the database itself is modified).

In the beginning of this chapter, we mentioned that one of the main reasons of using a database in web mapping is that we can display subsets of the data, filtered according to user input (Section 9.1). That way, we can have large amounts of data "behind" the web map, while maintaining responsiveness thanks to the fact that small portions of the data are transferred to the client each time. To fully exploit the advantages of connecting a database to a web map, in the next two Chapters 10–11 we will see examples where the SQL query is generated *dynamically*, in response to user input:

- In Chapter 10, we will load data according to an *attribute* value the user selects in a dropdown menu.
- In Chapter 11, we will load data according to *spatial* proximity to a clicked location on the map.

9.9   Exercise

- The following SQL query returns the (sorted) species list from the `plants` table: `SELECT DISTINCT name_lat FROM plants ORDER BY name_lat` (see Section 10.4.3).
- Load the result of the query inside a web page, and use it to dynamically generate an unordered list (`<ul>`) of all unique plant species names in the `plants` table (Figure 9.8).
- Since the result of the query is non-spatial, as it does not contain the geometry column, you need to use `format=JSON` in the query URL.
- Hint: use `example-04-08.html` from Section 4.13 (Figure 4.9), where we generated an unordered list based on an array, as a starting point for this exercise.

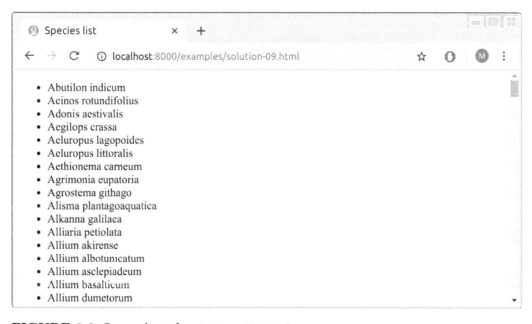

FIGURE 9.8: Screenshot of `solution-09.html`

10

Non-spatial Queries

10.1 Introduction

In Chapter 9, we introduced the CARTO SQL API, and used it to display the result of a database query on a Leaflet map (Figure 9.7). The query we used was constant: the same observations—namely the first 25 records in the **plants** table—were always loaded. In this chapter, we are going to extend **example-09-01.html** by adding a *user interface*, where one can choose which subset of the data gets loaded and displayed on the map. Specifically, we will use an input element, where the user selects a plant species, which we will use to dynamically generate an SQL query for filtering the **plants** table and getting the observations for that particular species only. That way, the user is able to load all observations of any given plant species, one at a time.

We will accomplish the above by going through several steps:

- Using SQL filtering based on column values (Section 10.2.1)
- Loading a subset of records from a database onto a Leaflet map (Section 10.2.2)
- Adding a **<select>** input for choosing a species (Section 10.3)
- Collecting all unique species names, also using SQL (Section 10.4.3)
- Automatically adding all unique species names as **<option>** elements inside **<select>** (Section 10.4.4)
- Associating the list with the map, so that changing the species choice will load new observations on the map (Section 10.5–10.6)

The final result we aim at is shown in Figure 10.7.

10.2 Subsetting with SQL

10.2.1 Filtering based on attributes

In Chapter 9, we used the following SQL query to load the first 25 records from the **plants** table on a Leaflet map (Figure 9.7).

```
SELECT name_lat, the_geom
  FROM plants
  LIMIT 25;
```

As mentioned in Section 9.6.2, using SQL we can also *filter* the returned data, using the `WHERE` keyword combined with one or more criteria. For example, the `name_lat` column specifies the Latin name of the observed species in the `plants` table. In case we want to select the observations of just one species, we can add the `WHERE name_lat = '...'` expression to our SQL query, where ... is replaced with the species name of interest. For example, the following query returns all observations where `name_lat` is equal to `'Iris mariae'` (Figure 10.1):

```
SELECT name_lat, the_geom
  FROM plants
  WHERE name_lat = 'Iris mariae';
```

For simplicity, let's add `LIMIT 1` to examine just the first observation for `'Iris mariae'`:

```
SELECT name_lat, the_geom
  FROM plants
  WHERE name_lat = 'Iris mariae'
  LIMIT 1;
```

When working with CARTO, we can execute this query with the corresponding SQL API call (Section 9.7):

```
https://michaeldorman.carto.com/api/v2/sql?format=GeoJSON&q=
SELECT name_lat, the_geom FROM plants WHERE name_lat = 'Iris mariae' LIMIT 1
```

Here is the returned GeoJSON string:

```
{
  "type": "FeatureCollection",
  "features": [
    {
      "type": "Feature",
      "geometry": {
        "type": "Point",
        "coordinates": [34.412502, 31.019879]
      },
      "properties": {
        "name_lat": "Iris mariae"
      }
    }
  ]
}
```

Indeed, this is a `"FeatureCollection"` (Section 7.3.4) with one feature and one non-spatial attribute (`name_lat`), representing a single record for `'Iris mariae'` as requested from the server.

FIGURE 10.1: *Iris mariae*, a rare iris species found in Israel. The image was taken in an ecological greenhouse experiment at the Ben-Gurion University in 2008.

10.2.2 Displaying on a map

For adding the observations of a single species on a Leaflet map, we can take `example-09-01.html` from Section 9.8 and modify just the SQL query part. Instead of the query we used in `example-09-01.html`, we can use the new one that we have just constructed, where an individual species is being selected:

```
SELECT name_lat, the_geom
  FROM plants
  WHERE name_lat = 'Iris mariae';
```

After creating the `map` object and loading a tile layer, the following code can be inserted to load the observations for `'Iris mariae'` on the map, along with popups:

```
var urlGeoJSON = "https://michaeldorman.carto.com/api/v2/sql?" +
    "format=GeoJSON&q=";
var sqlQuery1 = "SELECT name_lat, the_geom FROM plants " +
    "WHERE name_lat = 'Iris mariae'";
$.getJSON(urlGeoJSON + sqlQuery1, function(data) {
    L.geoJSON(data, {
        onEachFeature: function(feature, layer) {
            layer.bindPopup("<i>" + feature.properties.name_lat + "</i>");
        }
    }).addTo(map);
});
```

This code section uses the `$.getJSON` function for loading GeoJSON with Ajax (Section 7.7.3), as well as the `onEachFeature` option of `L.geoJSON` to bind specific popups per feature (Section 8.5). The popups display the `name_lat` property, in *italic* font. As discussed in Section 9.8, we are requesting the GeoJSON content from the CARTO SQL API, using a URL for a specific SQL query, specified in `sqlQuery1`.

The resulting map (`example-10-01.html`), with marked `'Iris mariae'` observations, is shown in Figure 10.2.

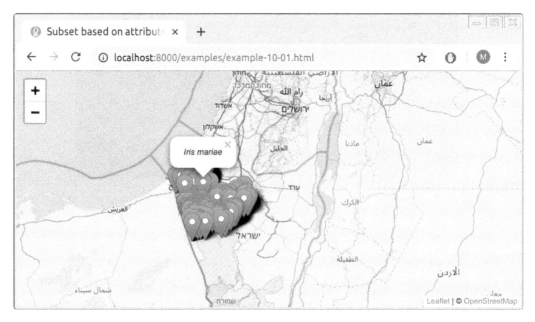

FIGURE 10.2: Screenshot of `example-10-01.html`

10.3   Creating a dropdown menu

Our ultimate goal in this chapter is for the user to be able to *choose* a species he/she wants to display, rather than seeing a predefined species such as `'Iris mariae'` (Figure 10.7). Therefore, our next step is to add an input element for selecting a species to load on the map. Since there are several hundreds of species in the `plants` table[1], a bulleted list (Section 1.6.7) or a set of radio buttons (Section 1.6.12.6) will be too long to fit on screen and inconvenient to browse. A **dropdown menu** (Section 1.6.12.8) is a good choice in case there are a lot of options to choose from. A dropdown menu can be opened to browse the species list and automatically closed again when selection is made, thus only temporarily obstructing other content on the web page.

As discussed in Section 1.6.12.8, a dropdown menu can be constructed using the `<select>` element, with an internal set of `<option>` elements for the various options. For example, suppose that we had just three plant species in the database, named `"Abutilon indicum"`,

---

[1]There are 417 species in the `plants` table, to be exact. See how to find that out in Section 10.4.1.

`"Acinos rotundifolius"`, and `"Adonis aestivalis"`. In such case, the dropdown menu could be defined with HTML code such as follows:

```
<select id="species_sel">
    <option value="Abutilon indicum">Abutilon indicum</option>
    <option value="Acinos rotundifolius">Acinos rotundifolius</option>
    <option value="Adonis aestivalis">Adonis aestivalis</option>
</select>
```

Each `<option>` element has a `value` attribute, as well as text contents. Recall from Section 1.6.12 that the `value` attribute specifies the ID sent to the server, or queried with the `.val` method of jQuery (Section 4.7.7), while the text contents between the opening and closing tags (`<option>...</option>`) is actually displayed on screen. In this particular example, the `value` attribute and the text content are identical, but this does not necessarily have to be so (Section 1.6.12). In fact, when the `value` attribute is missing, the element contents are treated as the values[2]. Therefore, when text contents and `value` are identical, the value can be omitted, as in `<option>Abutilon indicum</option>`. The `<select>` element can thus be defined with more concise code, as follows:

```
<select id="species_sel">
    <option>Abutilon indicum</option>
    <option>Acinos rotundifolius</option>
    <option>Adonis aestivalis</option>
</select>
```

To incorporate a dropdown menu—such as the one coded above—into our Leaflet map, we use the `L.control` function, which we already know how to use from Sections 6.8, 8.6, and 8.8.2. The following code can be added to our `<script>` to build the dropdown menu:

```
var dropdown = L.control({position: "topleft"});
dropdown.onAdd = function(map) {
    var div = L.DomUtil.create("div", "dropdown");
    div.innerHTML =
        '<select id="species_sel">' +
            '<option>Abutilon indicum</option>' +
            '<option>Acinos rotundifolius</option>' +
            '<option>Adonis aestivalis</option>' +
        '</select>';
    return div;
};
dropdown.addTo(map);
```

Like in previous examples of `L.control`, the above code is actually composed of three expressions:

- **Defining** a map control object, named `dropdown`, using `L.control`.
- **Setting** the contents of the control, assigning custom HTML code into the `.innterHTML` property. Note that the control itself is set as a `<div>` element with class `"dropdown"`, and the `<select>` element inside the `<div>` is set with an ID attribute of `"species_sel"`.
- **Adding** the `dropdown` control to our map.

---

[2] `https://developer.mozilla.org/en-US/docs/Web/HTML/Element/option`

Next, we probably want to improve the default styling of the dropdown menu, by adding some CSS rules applied to `#species_sel`, which refers to the `<select>` input element. For example, we can make the font slightly larger and add shadow effects around the input box. To do that, we add the following CSS code inside the `<style>` element within the `<head>`:

```
#species_sel {
    padding: 6px 8px;
    font: 14px/16px Arial, Helvetica, sans-serif;
    background-color: rgba(255,255,255,0.8);
    box-shadow: 0 0 15px rgba(0,0,0,0.2);
    border-radius: 5px;
}
```

Here is a short explanation of what each of the above styling rules does. You can go back to Chapter 2 for more information on each of these properties.

- `padding`—The padding clears an area around the contents (and inside the border) of an element. When two values are given, the first one (`6px`) refers to `padding-top` and `padding-bottom` and the second one (`8px`) refers to `padding-right` and `padding-left` (Section 2.8.4).
- `font`—Specifies the font properties. When two sizes are given, the first one (`14px`) refers to `font-size`, and the second one (`16px`) refers to `line-height` (the total height of text plus the distance to the next line). The font specification first gives specific fonts (`Arial`, `Helvetica`), then a general font family (`sans-serif`) to fall back on if those fonts are not installed on the system (Section 2.8.3).
- `background-color`—The box background color is given in the RGBA format: intensity of red, green, blue, and transparency (alpha). In this case, we have white at 80% opacity (`rgba(255,255,255,0.8)`) (Section 2.8.2).
- `box-shadow`—Adds a shadow around the element's box. The sizes refer to horizontal offset (`0`) and vertical offset (`0`), where zero means the shadow is symmetrical around all sides of the box. Blur distance (`15 px`) determines how far the shadow extends. Color specification comes next, in this case the shadow color is black at 20% opacity (`rgba(0,0,0,0.2)`).
- `border-radius`—Adds rounded corners to an element, the size (`5px`) sets the radius (Section 2.8.4.3).

The reason we do not need any styling rules for the entire control `<div>` (using the class selector `.dropdown`), just for the internal `<select>` element (using the ID selector `#species_sel`), is that the control contains nothing but the dropdown menu. The exercise at the end of this chapter (Section 10.7) will require creating a more complex control, with two dropdown menus. In that case, styling the entire control `<div>` is necessary too—for example, setting its external border style and the style of text labels for the separate dropdown menus (Figure 10.8).

- Open `example-10-02.html`.
- Uncheck each of the dropdown menu CSS styling rules in the developer tools, or delete them from the source code, to see their effect on dropdown menu appearance.

The resulting map `example-10-02.html` is shown in Figure 10.3. Note the newly added dropdown menu in the top-left corner of the map.

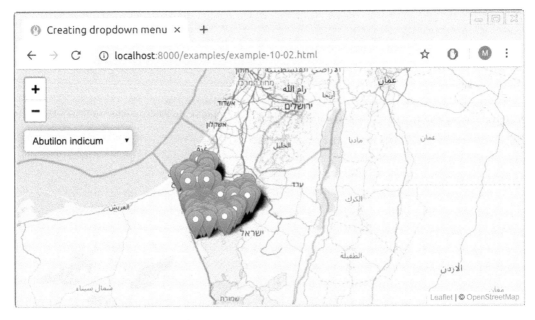

FIGURE 10.3: Screenshot of `example-10-02.html`

The dropdown menu with the three individual species—which we see in the top-right corner in Figure 10.3—is not yet functional. We can make a selection of a different species, but the map will still show `'Iris mariae'` nevertheless, since it is hard-coded into our SQL query `sqlQuery1` (open `example-10-02.html` and try it).

Before making the necessary adjustments for linking our dropdown menu with the marker layer on the map, we would like to improve the way that we create the dropdown menu in the first place. Instead of using pre-defined HTML code with just three species (as in `example-10-02.html`), we would like to build the options dynamically, with JavaScript code. Remember `example-04-08.html` (Section 4.13), where we used a `$.each` iteration (Section 4.12), for dynamically building a list of species names based on an array? This is exactly the technique we will use for dynamically populating our species dropdown menu with all of the unique species names.

10.4 Populating dropdown options

10.4.1 How many species do we have?

Before looking into the dropdown menu construction, let's check how many different plant species we actually have in the `plants` table. The following SQL query can be used to do that:

```
SELECT COUNT(DISTINCT name_lat) AS n
  FROM plants;
```

We are using the COUNT function along with the DISTINCT keyword to get the number of unique values in the name_lat column, putting the result into a new column named n. The result is a small table with one column (n) and one row, as follows:

```
  n
-----
 417
(1 row)
```

We can get this result with the CARTO SQL API too, as follows:

```
https://michaeldorman.carto.com/api/v2/sql?q=
SELECT COUNT(DISTINCT name_lat) AS n FROM plants
```

Here is the response:

```
{
  "rows": [
    {"n": 417}
  ],
  "time": 0.04,
  "fields": {
    "n": {
      "type": "number"
    }
  },
  "total_rows": 1
}
```

Omitting the format parameter implies the default format=JSON, thus our result is a JSON object (Section 9.7.2). In JSON results coming from the CARTO SQL API, the "rows" property has the actual tabular content. The other properties in the JSON object are *metadata*, with information regarding processing time ("time"), the column data types ("fields"), and the total number of rows ("total_rows")[3].

The "rows" property is structured as an array, with array elements representing table rows. Each element is an object, representing all of the column values for that table row. In this case, there is just one row and one column, which is why the array has just one element, and that element is an object with just one property:

```
[{"n": 417}]
```

The response tells us that there are 417 species in the plants table, which we would like to include as dropdown menu options. Manually writing the <option> elements code for each and every one of the species is obviously not a good idea (Section 4.13). Automating the construction of HTML code with 417 <option> element will be more convenient. What is more important, however, is the fact that our dynamically generated dropdown menu will be based on the exact species present in the database, in real-time. In other words, with the dynamic approach we no longer need to worry about updating the HTML code

---

[3]Remember that, as mentioned in Section 9.7.2, applying a non-spatial query (such as the unique species count) combined with the format=GeoJSON option will result in an error, because in order to produce a GeoJSON object the query must include the geometry column.

in case the database changes, since the HTML code is dynamically constructed based on the database itself and therefore always up-to-date. Remember that unlike a local file, the CARTO database, just like any other database, may be accessed and modified by different users on different platforms (Section 9.1). For example, a database with a list of rare and protected species may be periodically updated by the GIS department of the Nature and Parks authority, adding new observations, removing species that are no longer considered rare, and so on. If our web page connects to the database to dynamically generate its contents on page load, we do not need to worry about changing the code to reflect database modifications; the up-to-date version will be automatically loaded each time[4].

- Removing the `DISTINCT` keyword from the above query gives the overall number of values in the `name_lat` column, that is, the total number of records in the table.
- Execute the query through the CARTO SQL API to find out how many individual plant observations are there in the `plants` table[5] (also see Section 12.6).

10.4.2 Dropdown menu placeholder

We now begin with populating the dropdown options. As a first step, it is convenient to create an empty placeholder, like we did in the species list example from Section 4.13. That way, the placement of the contents on our page is easier to manage. To create the empty placeholder, instead of the HTML code for a `<select>` element with three pre-defined `<option>` elements (Section 10.3), we place the following HTML code which creates an *empty* `<select>` element, with no internal `<option>` elements at all:

```
var dropdown = L.control({position: "topright"});
dropdown.onAdd = function(map) {
    var div = L.DomUtil.create("div", "dropdown");
    div.innerHTML = '<select id="species_sel"></select>';
    return div;
};
dropdown.addTo(map);
```

Just like in `example-10-02.html` (Section 10.3), we hereby set an ID `"species_sel"` for the `<select>` element. We already used the ID for styling the dropdown menu with CSS (Section 10.3). Now, in `example-10-03.html` (Section 10.4.4), we will also use the ID in our `<script>` to dynamically populate the dropdown `<option>` elements. To do that, we need to:

- Get the list of all unique species in the database (Section 10.4.3)
- Iterate over the species, adding an `<option>` element inside the `<select>` for each one of them (Section 10.4.4)

---

[4]Note that the particular CARTO database used in this example is for demonstration, not directly related to the Nature and Parks authority, or any other users, and thus remains unchanged, but the principle still holds.

[5]The answer is 23,827.

10.4.3   Finding unique values

The following SQL query—which we already mentioned in the last exercise (Section 9.9)—can be used to get the list of all unique values in the name_lat column of the plants table:

```
SELECT DISTINCT name_lat
  FROM plants
  ORDER BY name_lat;
```

The SELECT DISTINCT keyword combination before the name_lat column name ensures that we only get the unique values in that column. We are also using the ORDER BY keyword to have the resulting table of species names **sorted** by alphabetical order. It makes sense that the options in the dropdown menu are alphabetically ordered: that way the user can easily locate the species he/she is interested in, by scrolling towards that species and/or by typing its first letter with the keyboard. Also, all species of the same *genus* will be listed together, since they start with the same word.

You can test the above query with the CARTO API:

```
https://michaeldorman.carto.com/api/v2/sql?q=
SELECT DISTINCT name_lat FROM plants ORDER BY name_lat
```

The JSON response should look as follows[6]:

```
{
  "rows": [
    {"name_lat":"Abutilon indicum"},
    {"name_lat":"Acinos rotundifolius"},
    {"name_lat":"Adonis aestivalis"},
    ...
  ]
  "time": 0.032,
  "fields": {
    "name_lat": {
      "type": "string"
    }
  },
  "total_rows": 417
}
```

Next (Section 10.4.4), we will use this JSON response to fill up the dropdown menu where the user selects a species to display on the map.

10.4.4   Adding the options

To populate the dropdown menu, we need to incorporate the above SELECT DISTINCT SQL query (Section 10.4.3) into our script. First, we define a variable named sqlQuery2 for the

---

[6]The "rows" property, which contains the returned table, is truncated here to save space, with the ... symbol representing further objects for the other 414 species not shown.

latter SQL query, listing the distinct species. At this point, we thus have two different SQL queries in our script:

- `sqlQuery1` for loading observations of *one* species (Section 10.2.2)
- `sqlQuery2` for finding all *unique* species names

For now, in `sqlQuery1` we manually set the first displayed species to `'Abutilon indicum'`, since it is the first species in terms of alphabetic ordering, as shown in the above SQL API response (Section 10.4.3). The species named `'Abutilon indicum'` therefore comprises the first option in our dropdown menu. Later on, we will replace this *manual* way of selecting the first species with an automatic one, without hard-coding its name into the script (Section 10.6).

Also note that we have two different URL prefixes:

- `urlJSON` for getting `sqlQuery2` in JSON format, using the `format=JSON` option (unspecified, since it is the default)
- `urlGeoJSON` for getting `sqlQuery1` in GeoJSON format, using the `format=GeoJSON` option

Here are the definitions of both URL prefixes and queries:

```
// URL prefixes
var url = "https://michaeldorman.carto.com/api/v2/sql?";
var urlJSON = url + "q=";
var urlGeoJSON = url + "format=GeoJSON&q=";

// Queries
var sqlQuery1 = "SELECT name_lat, the_geom FROM plants " +
    "WHERE name_lat = 'Abutilon indicum'";
var sqlQuery2 = "SELECT DISTINCT name_lat FROM plants ORDER BY name_lat";
```

Then, the following expression fills the dropdown menu with all unique species names in the `plants` table, using `sqlQuery2`:

```
$.getJSON(urlJSON + sqlQuery2, function(data) {
    $.each(data.rows, function(key, value) {
        $("#species_sel").append("<option>" + value.name_lat + "</option>");
    });
});
```

This is an Ajax request using `$.getJSON`. What does it do? Let's go over the code step by step.

- The `$.getJSON` function is used to load the JSON response with the unique species names from the database, using `urlJSON+sqlQuery2`, passed as an object named `data`. The contents of `data` is the same as shown in the JSON printout from Section 10.4.3.
- Then, the `$.each` function (Section 4.12) is used to iterate over each element in the `rows` property of `data` (i.e., the returned table rows), applying a function that:
  - locates the `#species_sel` selector, referring to the dropdown menu; and
  - appends (Section 4.7.4) an internal `<option>` element before the closing tag of the `<select>` element, with the HTML element contents being the `name_lat` property of the current row value.

Note that, in the last code section, we are actually searching the DOM for the dropdown menu reference "#species_sel" 417 times, since the expression $("#species_sel") is executed *inside* the iteration. This is not necessary. We can locate the "#species_sel" element just once, keeping its reference in a variable, such as the one named menu in the alternative code section shown below. That way, the browser searches for the #species_sel reference only once, and uses that same reference inside the iteration. The performance difference is negligible in this case, but this is an example of an optimization: getting rid of actions that are unnecessarily repeated in our code:

```
$.getJSON(urlJSON + sqlQuery2, function(data) {
    var menu = $("#species_sel");
    $.each(data.rows, function(key, value) {
        menu.append("<option>" + value.name_lat + "</option>");
    });
});
```

As a result of adding either one of the last two code sections, we now have a map (example-10-03.html) with a dropdown menu listing all available species. The observations of the first species on the list are shown on the map on page load (Figure 10.4).

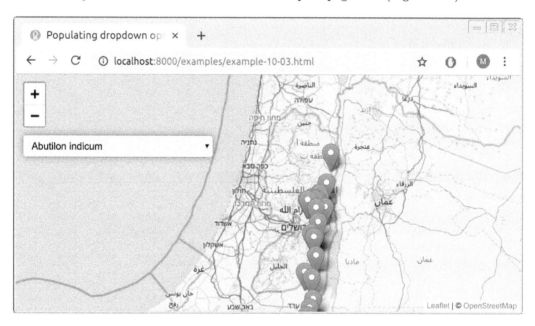

FIGURE 10.4: Screenshot of example-10-03.html

- Insert a console.log expression inside the <script> of example-10-03.html so that the name of each species is printed in the console as it is being added to the list of options.
- You should see a printout with species names (Figure 10.5).

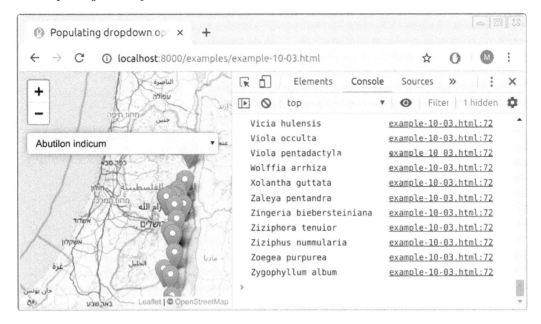

FIGURE 10.5: Species names printed with `console.log`

10.5 Updating the map

10.5.1 Overview

In `example-10-03.html`, we now have a web map showing observations of the *first* species from the `plants` table, as well as a dropdown menu with *all* species that are present in the `plants` table (Figure 10.4). Try clicking on the dropdown menu in `example-10-03.html` and selecting a different species. Nothing happens! What is still missing is an event listener, triggering an update of the plant observations layer on selection change. In our case, whenever the species selection changes, the event listener function should do the following things:

- **Clear** the map of all previously loaded observations
- **Define** a new SQL query for getting the observations of the currently selected plant species
- **Send** the SQL query to the database and get the GeoJSON with the new observations
- **Display** the new observations on the map

10.5.2 Manual example

Before setting up the event listener, it is often convenient to start with a manually defined, specific case. Let us suppose the user chose a different species instead of `'Abutilon indicum'`, say `'Iris mariae'`, in the dropdown menu. The code for adding `'Iris mariae'` observations on the map is exactly the same as the one we used for `'Abutilon indicum'`. The only thing that's changed is the species name in the SQL query, hereby named `sqlQuery3`.

Try entering the following code section into the console of `example-10-03.html` (Figure 10.4). This should load new observations, for `'Iris mariae'`, on top of the existing ones for `'Abutilon indicum'`:

```javascript
var valueSelected = "Iris mariae";
var sqlQuery3 = "SELECT name_lat, the_geom FROM plants " +
    "WHERE name_lat = '" + valueSelected + "'";
$.getJSON(urlGeoJSON + sqlQuery3, function(data) {
    L.geoJSON(data, {
        onEachFeature: function(feature, layer) {
            layer.bindPopup("<i>" + feature.properties.name_lat + "</i>");
        }
    }).addTo(map);
});
```

- Change the value of `valueSelected` in the above code section to a different species name, such as `"Iris atropurpurea"`, then execute the modified code section in the console of `example-10-03.html`.
- You should see observations of the species you chose added on the map.

10.5.3 Automatic updating

In the last code section, we manually added new observations of another species from the database. What we now need is to associate this code section with dropdown menu *changes*. In other words, we want the selection of a new species in the menu to trigger the automatic removal of the old observations and loading the new ones. This is where the event listener comes into play.

The first thing we do is define a `layerGroup` named `plantsLayer` and add it on the map:

```javascript
var plantsLayer = L.layerGroup().addTo(map);
```

The named layer group can be referenced later on, to clear it or add layers into it. This will make it easier for us to clear any previously loaded species and add markers for the currently selected one. Recall the GeoJSON viewer in `example-07-02.html` (Section 7.6), where we used a layer group for clearing old GeoJSON layers whenever a new one is submitted from the text area input.

Accordingly, we replace the `.addTo(map)` part with `.addTo(plantsLayer)` when loading the initial species on the map. So, the following expression[7] from `example-10-03.html`:

```javascript
$.getJSON(urlGeoJSON + sqlQuery1, function(data) {
    L.geoJSON(...).addTo(map);
});
```

---

[7]Note that the arguments passed to `L.geoJSON` were replaced with ... to save space.

is replaced with this one in **example-10-04.html**:

```
$.getJSON(urlGeoJSON + sqlQuery1, function(data) {
    L.geoJSON(...).addTo(plantsLayer);
});
```

The second thing we do is "wrap" the manual example from the previous section, which we used to load **"Iris mariae"** observations, so that:

- The code for loading observations on the map is contained inside an event listener function, responding to changes in the dropdown menu (**"#species_sel"**). The appropriate event type in this case is **"change"** (Section 4.3.4), which means that any *change* in the input element triggers the event listener.
- The species to load from the database is determined based on the currently selected value in the dropdown menu, using **$("#species_sel").val()**, rather than hard coded as **"Iris mariae"**.
- The old observations are cleared before adding new ones, using **.clearLayers()**.

The following code section does all of those things:

```
$("#species_sel").on("change", function() {
    var valueSelected = $("#species_sel").val();
    var sqlQuery3 = "SELECT name_lat, the_geom FROM plants " +
        "WHERE name_lat = '" + valueSelected + "'";
    $.getJSON(urlGeoJSON + sqlQuery3, function(data) {
        plantsLayer.clearLayers();
        L.geoJSON(data, {
            onEachFeature: function(feature, layer) {
                layer.bindPopup(
                    "<i>" + feature.properties.name_lat + "</i>"
                );
            }
        }).addTo(plantsLayer);
    });
});
```

Let's go over the code, step by step. In the outermost part of the expression, an event listener is being binded to a DOM selection:

```
$("#species_sel").on("change", function() {
    ...
});
```

More specifically, an anonymous function is being set to execute every time the input element **"#species_sel"** changes. Let's now review the body of that anonymous function. First thing, the function defines a local variable named **valueSelected**:

```
var valueSelected = $("#species_sel").val();
```

The value of `valueSelected` is the name of the currently selected species in the dropdown menu. The current value is obtained with the `.val` method (Section 4.7.7). This is conceptually similar to the calculator example `example-04-09.html` (Figure 4.10), where we used `.val` to get the currently entered numbers, for displaying their multiplication product (Section 4.14).

Then, `valueSelected` is being used to construct a new SQL query string `sqlQuery3`. This query will be used to request the observations of the newly selected species from the database:

```
var sqlQuery3 = "SELECT name_lat, the_geom FROM plants " +
    "WHERE name_lat = '" + valueSelected + "'";
```

Finally, the new GeoJSON content is being requested with `urlGeoJSON+sqlQuery3` using the `$.getJSON` function. Once the request finishes, the returned GeoJSON object is added on the map using `L.geoJSON` and `.addTo(plantsLayer)`, but not before the old observations are cleared with `plantsLayer.clearLayers()`:

```
$.getJSON(urlGeoJSON + sqlQuery3, function(data) {
    plantsLayer.clearLayers();
    L.geoJSON(data, {
        onEachFeature: function(feature, layer) {
            layer.bindPopup("<i>" + feature.properties.name_lat + "</i>");
        }
    }).addTo(plantsLayer);
});
```

The resulting map (`example-10-04.html`) is shown in Figure 10.6. Thanks to the event listener, the dropdown menu is now functional, and the map is responsive to the current species selection.

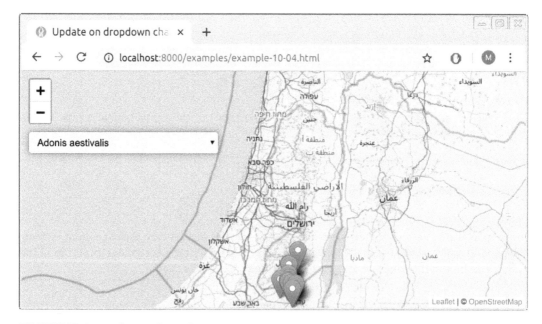

FIGURE 10.6: Screenshot of `example-10-04.html`

10.6   Refactoring the code

Going over the complete code of `example-10-04.html` (Section 10.5.3), you may notice one important drawback: there are two code sections doing practically the same thing. Namely, we have the following code section for loading the initial query `sqlQuery1`, displaying the *initial* species `'Abutilon indicum'` on page load:

```
var sqlQuery1 = "SELECT name_lat, the_geom FROM plants" +
    "WHERE name_lat = 'Abutilon indicum'";
$.getJSON(urlGeoJSON + sqlQuery1, function(data) {
    L.geoJSON(...).addTo(plantsLayer);
});
```

and we also have the event listener, which does the same thing for each *subsequently* selected species, using a dynamically generated query `sqlQuery3`[8]:

```
$("#species_sel").on("change", function() {
    var valueSelected = $("#species_sel").val();
    var sqlQuery3 = "SELECT name_lat, the_geom FROM plants " +
        "WHERE name_lat = '" + valueSelected + "'";
    $.getJSON(urlGeoJSON + sqlQuery3, function(data) {
        plantsLayer.clearLayers();
        L.geoJSON(...).addTo(plantsLayer);
    });
});
```

This duplication contradicts the **Don't Repeat Yourself**[9] (DRY) principle—an essential guideline in computer programming. Basically, unnecessary duplication makes our code more difficult to manage, since we need to keep track of all duplicated parts and make sure they are synchronized every time we make changes in our code. For example, suppose we decide to change the popup content for the plant observation markers, say, displaying the observation date below the species name. In the repeated code structure from `example-10-04.html`, we would have to change `L.geoJSON` options in two places: the `L.geoJSON` function call that loads the initial species and the `L.geoJSON` function call that loads any subsequently selected species. If we forget to change one of those places, then popup content will look different depending on whether we changed the initial dropdown menu selection, which is clearly undesired.

A second issue in the present code configuration is that the initial species, the one displayed on map load, is hard-coded in `sqlQuery1` (it is set to `'Abutilon indicum'`). Like we said previously (Section 10.4.1), the database may change in the future, so that `'Abutilon indicum'` may be removed or a new species with alphabetical precedence may be added. In such a case, the initial view will be incorrect: the species initially loaded on the map will no longer match the first species inside the dropdown menu. Therefore, it is beneficial to determine the first species to load according to the real-time version of the database, rather than based on a constant string.

---

[8] Again, note that the arguments passed to `L.geoJSON` were replaced with ... to save space.
[9] https://en.wikipedia.org/wiki/Don%27t_repeat_yourself

The solution to the duplication issue is to use a **function**, which also makes our code slightly shorter and more elegant. Instead of repeating the code section for loading plant observations in two places, we wrap the second code section—the one for loading the currently selected species—in a function called displaySpecies. The displaySpecies function loads the currently selected species on the map:

```
function displaySpecies() {
    var valueSelected = $("#species_sel").val();
    var sqlQuery3 = "SELECT name_lat, the_geom FROM plants " +
        "WHERE name_lat = '" + valueSelected + "'";
    $.getJSON(urlGeoJSON + sqlQuery3, function(data) {
        plantsLayer.clearLayers();
        L.geoJSON(data, {
            onEachFeature: function(feature, layer) {
                layer.bindPopup(
                    "<i>" + feature.properties.name_lat + "</i>"
                );
            }
        }).addTo(plantsLayer);
    });
}
```

The displaySpecies function can be used in an .on("change", ...) event listener, just like the anonymous function was in the previous version, so the event listener definition from example-10-04.html:

```
$("#species_sel").on("change", function() {...});
```

is replaced with the following version in example-10-05.html:

```
$("#species_sel").on("change", displaySpecies);
```

This change takes care of clearing the map and loading a new species whenever the dropdown menu selection changes. However, how will the initial species be loaded and displayed on map load? To load the initial species, we simply need to call the function one more time in our script, *outside* of the event listener. The appropriate place to call the function is inside the Ajax request that populates the dropdown menu. That way, the first species is displayed right after the dropdown menu is filled with all of the unique species names:

```
$.getJSON(urlJSON + sqlQuery2, function(data) {
    $.each(data.rows, function(key, value) {
        $("#species_sel").append("<option>" + value.name_lat + "</option>");
    });
    displaySpecies(); // Display initial species
});
```

Since displaySpecies is being called after the dropdown menu was already populated, we can be sure that the initially loaded species corresponds to the first selected <option>, even if the database has changed. We no longer need to worry that a particular fixed species name in our code (`'Abutilon indicum'`) still matches the first one in the alphabetically ordered species list from the database.

The resulting map `example-10-05.html` is shown in Figure 10.7. Visually and functionally it is exactly the same as `example-10-04.html`. However, the underlying code is rearranged and improved, which is also known as **code refactoring**[10].

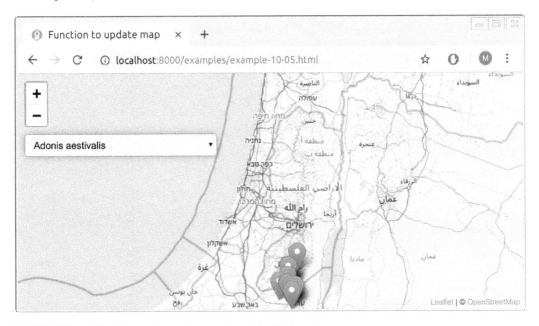

FIGURE 10.7: Screenshot of `example-10-05.html`

10.7 Exercise

- Extend `example-10-05.html` into a web map with two dropdown menus—one for choosing the *genus* and another one for choosing the *species* (Figure 10.8).
- When the user selects a genus, the second dropdown should automatically update to list the species in that genus. The user then selects a species to show it on the map. If the user made no selection yet, the first species in each genus should be automatically displayed on the map.
- This is a challenging exercise, so here is a hint on how to start. Run the code section shown below in the console of `example-10-05.html`. This code constructs an object named `species`, whose properties are genus names and values are arrays with all species in that genus. Use the `species` object to populate the dropdown menus.

---

[10]https://en.wikipedia.org/wiki/Code_refactoring

```
var sqlQuery = "SELECT DISTINCT name_lat FROM plants ORDER BY name_lat";
var species = {};
$.getJSON(urlJSON + sqlQuery2, function(data) {
    $.each(data.rows, function(key, value) {
        var tmp = value.name_lat;
        var tmp_split = tmp.split(" ");
        if(species[tmp_split[0]] === undefined) {
            species[tmp_split[0]] = [tmp];
            } else {
                species[tmp_split[0]].push(tmp);
            }
    });
});
```

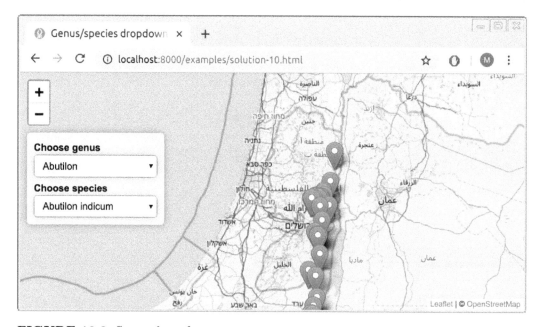

FIGURE 10.8: Screenshot of `solution-10.html`

11

Spatial Queries

11.1 Introduction

In this chapter, we are going to integrate our plant observations web map with spatial SQL queries to selectively load data from the CARTO database, based on geographical proximity. Specifically, we will build a web map where the user can click on any location, and as a result the nearest n plants observations to the clicked location will be loaded from the database and displayed, along with straight-line paths towards those plants. You can see the final result in Figure 11.8.

We will build the web map going through several steps, each time adding more functionality:

- Adding a marker at the clicked location on the map (Section 11.2.1)
- Adding a *custom* marker at clicked location, to distinguish it from the observation markers (Section 11.2.2)
- Finding n nearest features, using a spatial SQL query, and adding them on the map (Section 11.4)
- Drawing line segments from clicked location to the nearest features (Section 11.5)

11.2 Adding markers on click

11.2.1 Getting click coordinates

Our first step towards a web map that queries the **plants** table based on spatial proximity is to mark the clicked location on the map, while capturing its coordinates. The coordinates will be stored, to pass them on with a spatial SQL query (Section 9.6.4). We start from the basic map **example-06-02.html** from Section 6.5 and make changes on top of it. First, we modify the initially displayed map extent to a larger one:

```
var map = L.map("map").setView([32, 35], 8);
```

Next, we add a layer group named **myLocation**. This layer group will contain the clicked location marker, created on map click. As we have seen in Sections 7.6.4 and 10.5.3, using a layer group will make it easy to remove the old marker before adding a new one, in case the user changed his/her mind and subsequently clicked on a different location:

```
var myLocation = L.layerGroup().addTo(map);
```

We then define an event listener, which will execute a function named `mapClick` each time the user clicks on the map. Remember the map click event and its `.latlng` property introduced in the `example-06-08.html` (Section 6.9)? We use exactly the same principle here, only instead of adding a popup with the clicked coordinates in the clicked location, we are adding a *marker*. To do that, we first set the event listener for `"click"` events on the `map` object, referencing the `mapClick` function that we have yet to define:

```
map.on("click", mapClick);
```

Second, we define the `mapClick` function itself. The `mapClick` function needs to do two things:

- **Clear** the `myLocation` layer of any previous markers, from a previously clicked location (if any)
- **Add** a new marker to the `myLocation` layer, at the clicked coordinates, using the `.latlng` property of the event object

Here is the definition of the `mapClick` function:

```
function mapClick(e) {
    myLocation.clearLayers();
    L.marker(e.latlng).addTo(myLocation);
}
```

As a result, we now have a map (`example-11-01.html`) where the user can click, with the last clicked location displayed on the map (Figure 11.1).

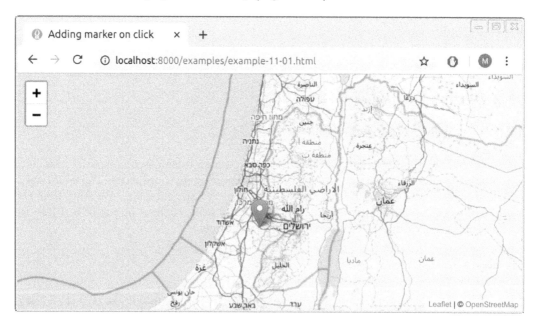

FIGURE 11.1: Screenshot of `example-11-01.html`

11.2.2 Adding a custom marker

Shortly, we will write code to load plant observations next to to our clicked location, and display them as markers too (Figure 11.7). To distinguish the marker for the clicked location from the markers for the plant observations, we need to override the default blue marker settings in Leaflet and display a different type of marker for one of the two layers. For example, we can use a *red* marker for the clicked location and keep using the default *blue* markers for denoting the plant locations (Figure 11.7). To change marker appearance, however, we first need to understand a little better how it is defined in the first place.

The markers we see on a Leaflet map, created with `L.marker` (Section 6.6.2), are in fact PNG images displayed at the specified locations placed on top of the map background. If you look closely, you will see that the default blue marker also has a "shadow" effect behind it, to create an illusion of depth, as if the marker "sticks out" of the map (Figure 11.2). The shadow is a PNG image too, displayed behind the PNG image of the marker itself. We can use the *Inspect Element* mode (Section 1.9) in the **developer tools** to figure out which PNG image is actually being loaded by Leaflet, and where it comes from (Figure 11.3). Doing so reveals the following values for the `src` attributes of the `<img>` elements for the marker and marker shadow images:

- `http://localhost:8000/examples/css/images/marker-icon-2x.png`
- `http://localhost:8000/examples/css/images/marker-shadow.png`

FIGURE 11.2: Default and custom images for drawing a Leaflet marker

Note that the prefix `http://localhost:8000/` is specific to a locally-served (Section 5.6.2) copy of the examples, and may be different when viewing them on a different server. The *local* Leaflet JavaScript library, which is included in `example-11-01.html`, looks for the marker PNG images in a sub-directory named `images` inside the directory where the `leaflet.css` file is placed (Section A). That is the reason for us placing the images in the sub-directory, which we mentioned in Section 6.5.3. In case we use a *remote* copy of the Leaflet library (Section 6.5.3), the markers would have been loaded from remote PNG files, such as the following ones:

- `https://unpkg.com/leaflet@1.5.1/dist/images/marker-icon-2x.png`
- `https://unpkg.com/leaflet@1.5.1/dist/images/marker-shadow.png`

You can follow these URLs to download and inspect the PNG images in any graphical viewer or editor, such as **Microsoft Paint**[1].

To distinguish our clicked location from other markers on the map, we will use a different marker for the clicked location. In our example, we will use a marker PNG image, which is similar to the default one, only colored in red instead of blue (Figure 11.2). The PNG

---

[1]`https://en.wikipedia.org/wiki/Microsoft_Paint`

FIGURE 11.3: Inspecting Leaflet icon `<img>` element

images for the red icon and its shadow are also included in the online book materials, in the `images` sub-directory (Appendix A):

- `images/redIcon.png`
- `images/marker-shadow.png`

To set a custom Leaflet marker, based on the PNG images `redIcon.png` and `marker-shadow.png`, you need to place these files on your server. For example, the files can be placed in a folder named `images` within the root directory of the web page (see Section 5.5.2), as given in the online materials. A custom marker using these PNG images, assuming they are in the `images` folder, is then defined with the `L.icon` function as follows:

```
var redIcon = L.icon({
    iconUrl: "images/redIcon.png",
    shadowUrl: "images/marker-shadow.png",
    iconAnchor: [13, 41]
});
```

The `L.icon` function requires a path to the PNG images for the icon (`iconUrl`), and, optionally, a path to the PNG image for the shadow (`shadowUrl`). The other important parameter is `iconAnchor`, which sets the anchor point, i.e., the exact icon image pixel which corresponds to the point coordinates where the marker is initiated. The custom icon object is assigned to a variable, hereby named `redIcon`, which we can later use to draw custom markers on the map with `L.marker`.

What is the meaning of `iconAnchor`, and why did we use the value of `[13, 41]`? The `iconAnchor` parameter specifies which pixel of the marker PNG image will be placed on the `[lon, lat]` point where the marker is initialized. To determine the right anchor point coordinates, we need to figure out the size (in pixels) of our particular marker, and the image region where we would like the anchor to be placed. The image size can be determined by viewing the PNG file properties, for example clicking on the file with the right mouse button, choosing **Properties** and then the **Details** tab. In our case, the size of the redIcon.png image is 25 by 41 pixels (Figure 11.4).

FIGURE 11.4: File properties for the red icon redIcon.png. The highlighted entry shows image dimensions in pixels (25 x 41).

Conventionally, image coordinates are set from a [0, 0] point at the top-left corner of the icon[2]. The anchor point for our particular icon should be at its tip, at the bottom-middle. Starting from the top-left corner [0, 0], this means going all the way *down* on the Y-axis, then half-way to the *right* on the X-axis. Since the PNG image size is [25, 41], this means we set the the pixel on the center of the X-axis ([13, ...]) and bottom of the Y-axis ([..., 41]), thus the anchor value of [13, 41] (Figure 11.5).

Now that the redIcon object is ready and the PNG images are in place, we can replace the expression for adding a marker inside the map click event listener in example-11-01.html:

```
L.marker(e.latlng).addTo(myLocation);
```

---

[2]Measuring the coordinates from top-left and using an inverted Y-axis (values increase when moving *down*) is an old convention in computer graphics, emerging in many different situations. Going back to example-04-07.html (Section 4.11), you will see that mouse coordinates in the browser window are also measured from the top-left corner.

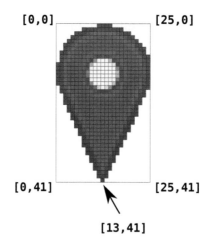

FIGURE 11.5: PNG image for the red marker `redIcon.png`, with coordinates of the four corners and the coordinate of the marker anchor point. Note that the coordinate system starts from the top-left corner, with reversed Y-axis direction.

with a new expression that loads our *custom* `redIcon` marker in `example-11-02.html`:

```
L.marker(e.latlng, {icon: redIcon}).addTo(myLocation);
```

The resulting map `example-11-02.html` is shown in Figure 11.6. Clicking on the map now adds the red marker icon instead of the default blue one.

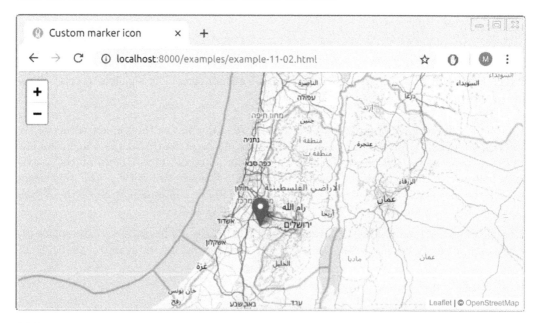

FIGURE 11.6: Screenshot of `example-11-02.html`

You can use just about any PNG image as a marker icon this way, not just a differently-colored default marker. There are many places where you can find PNG images suitable for map markers, such as the *Maps and Navigation* collection[3] on `https://www.flaticon.com` website[4].

- Download one of the PNG images for custom map icons from `https://www.flaticon.com/packs/maps-and-navigation-20`, or use any other small PNG image that you have.
- Adapt `example-11-02.html` to display your custom icon instead of the red icon.

11.3 Spatial PostGIS operators

11.3.1 Overview

Now that we know how to add a custom red marker on map click, we are going to add some code into our `mapClick` function incorporating a *spatial* SQL query (Section 9.6.4) to find the closest plants to the clicked location. As a result, each time the user clicks on a new point and `mapClick` adds a red marker, the function also queries for the nearest plants and displays them on the map (Figure 11.7).

As we have already discussed in Section 9.6.4, spatial databases are characterized by the fact that their tables may contain a geometry column. On CARTO, the geometry column is conventionally named `the_geom` (Section 9.7.2). The geometry column is composed of encoded geometric data in the WKB format, which can be used to transform returned table records to spatial formats, such as GeoJSON. For example, in Chapters 9–10 we used `SELECT` statements to return GeoJSON objects based on data from the `plants` table on CARTO. These queries utilized the `the_geom` geometry column to create the `"geomerty"` properties of the GeoJSON layer. However, the geometry column can also be used to make various spatial *calculations* on our data, using spatial SQL operators and functions. A very common example of a spatial calculation is the calculation of **geographical distance**.

In the next two sections (11.3.2–11.3.3), we will discuss the structure of the required spatial SQL query to return nearest features. Then, in Section 11.4, we will see how to implement this type of SQL query in our Leaflet web map.

11.3.2 Geographical distance

To find the five plant observations *nearest* to our clicked location, we can use the following query, which was introduced as an example of spatial SQL syntax in Section 9.6.4:

---

[3]`https://www.flaticon.com/packs/maps-and-navigation-20`

[4]For other examples of using custom icons in Leaflet, check out the *Leaflet Custom Markers* tutorial (`https://leafletjs.com/examples/custom-icons/`) and the *DUSPviz Map Design* tutorial (`http://duspviz.mit.edu/web-map-workshop/map-symbolization/`).

```
SELECT name_lat, obsr_date, ST_AsText(the_geom) AS geom
  FROM plants
  ORDER BY
    the_geom::geography <->
    ST_SetSRID(
      ST_MakePoint(34.810696, 31.895923), 4326
    )::geography
  LIMIT 5;
```

This query returns the nearest five observations from the **plants** table, based on distance to a specific point [34.810696, 31.895923]. In this section, we will explain the structure of the query more in depth.

First of all, as we already saw in Chapters 9–10, limiting the returned table to the first **n** records can be triggered using the term LIMIT **n**, as in LIMIT 5 or LIMIT 25. However, we need the five *nearest* plants, not just the five plants incidentally being first in terms of the table row ordering. This means the data need to be ordered before being returned.

As an example of *non-spatial* ordering, the table can be ordered with ORDER BY followed by column name(s) to select the five plant observations with the lowest (or highest) values for the given variable(s). We already saw one example of non-spatial ordering when producing the alphabetically ordered list of species in Section 10.4.3. As another example, consider the following query, which returns the five *earliest* plant observation records. The ORDER BY obsr_date part means the table is ordered based on the values in the obsr_date (observation date) column:

```
SELECT name_lat, obsr_date, ST_AsText(the_geom) AS geom
  FROM plants
  ORDER BY obsr_date
  LIMIT 5;
```

which gives the following result:

```
     name_lat     | obsr_date  |            geom
------------------+------------+----------------------------
 Iris haynei      | 1900-01-01 | POINT(35.654005 32.741372)
 Iris atrofusca   | 1900-01-01 | POINT(35.193367 31.44711)
 Iris atrofusca   | 1900-01-01 | POINT(35.189142 31.514754)
 Iris vartanii    | 1900-01-01 | POINT(35.139696 31.474149)
 Iris haynei      | 1900-01-01 | POINT(35.679761 32.770133)
(5 rows)
```

Try the corresponding CARTO SQL API query, choosing the CSV output format, to examine the above result on your own:

```
https://michaeldorman.carto.com/api/v2/sql?format=CSV&q=
SELECT name_lat, obsr_date, ST_AsText(the_geom) AS geom
FROM plants ORDER BY obsr_date LIMIT 5
```

Indeed, all returned records are from the 1900-01-01, which is the earliest date in the obsr_date field. *Spatial* ordering is similar, only that the table records are ordered based on their spatial arrangement, namely based on their distances to another spatial feature, or set of features. In other words, with spatial ordering, instead of ordering by non-spatial column values, we are ordering by geographical distances, which are calculated using the geometry column. In the above spatial query example for getting the five nearest points from a given location [34.810696, 31.895923], the only part different from the non-spatial query example is basically just the ORDER BY term. Instead of the following ORDER BY term for non-spatial ordering, based on obsr_date values:

```
ORDER BY obsr_date
```

we use the following ORDER BY term, for spatial ordering, based on distance from a specific point [34.810696, 31.895923]:

```
ORDER BY
  the_geom::geography <->
  ST_SetSRID(ST_MakePoint(34.810696, 31.895923), 4326)::geography
```

The result of the spatial query is as follows:

```
        name_lat       | obsr_date  |              geom
-----------------------+------------+----------------------------
 Lavandula stoechas    | 1931-04-30 | POINT(34.808564 31.897377)
 Bunium ferulaceum     |            | POINT(34.808504 31.897328)
 Bunium ferulaceum     | 1930-02-23 | POINT(34.808504 31.897328)
 Silene modesta        | 1900-01-01 | POINT(34.822295 31.900125)
 Corrigiola litoralis  | 2016-01-30 | POINT(34.825931 31.900792)
(5 rows)
```

These are the locations of the five nearest plants to the specified location. You can see that the longitude and latitude values are fairly similar to [34.810696, 31.895923], reflecting the fact that the points are proximate. Again, you can experiment with this query in the CARTO SQL API:

```
https://michaeldorman.carto.com/api/v2/sql?format=CSV&q=
SELECT name_lat, obsr_date, ST_AsText(the_geom) AS geom
FROM plants
ORDER BY the_geom::geography <-> ST_SetSRID(
ST_MakePoint(34.810696, 31.895923), 4326)::geography LIMIT 5
```

As you probably noticed, the expression used for spatial ordering is more complicated than simply a column name such as obsr_date:

```
the_geom::geography <->
ST_SetSRID(ST_MakePoint(34.810696, 31.895923), 4326)::geography
```

In addition to the geometry column name (the_geom), this particular ORDER BY term contains four spatial PostGIS functions and operations, which we will now explain:

- ST_MakePoint—Creates a point geometry
- ST_SetSRID—Sets the coordinate reference system (CRS)
- ::geography—Casts to the geography type
- <->—Calculates 2D distance

A two-dimensional point geometry is constructed using the ST_MakePoint function, given two coordinates x and y, in this case 34.810696 and 31.895923, respectively. Thus, the expression ST_MakePoint(34.810696, 31.895923) defines a single geometry of type "Point", which we can use in spatial calculations. The ST_SetSRID function then sets the coordinate reference system (CRS) for the geometry. The 4326 argument is the EPSG code of the WGS84 geographical projection (i.e., lon/lat) (Figure 6.1).

The ::geography part *casts*[5] the geometry to a special type called **geography**, thus determining that what follows is a calculation with **spherical geometry**, which is the appropriate way to do distance-based calculations with lon/lat data. With ::geography, distance calculations give the spherical **Great Circle**[6] distance, in meters (Figure 12.10). Omitting the ::geography part is equivalent to using the default ::geometry type, implying that what follows is a **planar geometry** calculation. Planar geometry calculations are only appropriate for projected data, which we do not use in this book. Using ::geometry when calculating distances on lon/lat data gives straight-line euclidean distances, in degree units, which is almost always inappropriate.

The <-> operator returns the 2D distance[7] between two sets of geometries. Since we set ::geography, the result represents spherical Great Circle distance in meters. In the present example, we are calculating the distances between **the_geom**, which is the geometry column of the **plants** table, and an individual point. The result is a series of distances in meters, corresponding to all features of the **plants** table.

Finally, the series of distances is passed to the ORDER BY keyword, thus rearranging the table based on the calculated distances, from the smallest to largest, i.e., from nearest observation to furthest. The LIMIT 5 part then takes the top five records, which are the five nearest ones to [34.810696, 31.895923].

11.3.3   Sphere vs. spheroid

As another demonstration of the four spatial PostGIS functions discussed in Section 11.3.2 (above), consider the following small query. This query calculates the distance between two points [0, 0] and [0, 1] in geographic coordinates (lon/lat), i.e., the length of one degree of longitude along the equator:

```
SELECT
  (ST_SetSRID(ST_MakePoint(0, 0), 4326)::geography <->
   ST_SetSRID(ST_MakePoint(1, 0), 4326)::geography) / 1000
AS dist_km;
```

According to the result, the distance is 111.195 km:

```
    dist_km
------------------
 111.195079734632
(1 row)
```

In this query, we are manually constructing two points in lon/lat, [0, 0] and [1, 0], using ST_MakePoint, ST_SetSRID and ::geography. The 2D distance operator <-> is the applied

---

[5]http://postgis.net/workshops/postgis-intro/geography.html
[6]https://en.wikipedia.org/wiki/Great_circle
[7]https://postgis.net/docs/geometry_distance_knn.html

on the points to calculate the Great Circle distance between them, in meters. Dividing the result by 1000 transforms the distance from meter to kilometer units[8].

The true distance between [0, 0] and [1, 0], however, is 111.320 km[9] and not 111.195. What is the reason for the discrepancy? The reason is that the <-> operator, though using spherical geometry, relies on a **sphere** model of the earth, rather than the more accurate **spheroid** model. In PostGIS, the more accurate but somewhat slower distance calculation based on a spheroid can be obtained with ST_Distance instead of <->, as in:

```
SELECT ST_distance(
  ST_SetSRID(ST_MakePoint(0, 0), 4326)::geography,
  ST_SetSRID(ST_MakePoint(1, 0), 4326)::geography) / 1000
AS dist_km;
```

This gives the more accurate result of 111.319 km:

```
     dist_km
-----------------
 111.31949079327
(1 row)
```

Though ST_distance gives the more accurate estimate, the calculation takes longer. For example, finding the 5 nearest neighbors from the plants table took 0.17 seconds using the <-> operator, compared to 0.37 seconds with ST_Distance, on an average laptop computer. This is a more than twice longer calculation time. Although, in this particular example, exactly the same five plants are returned in both cases, theoretically the ordering may differ among the two methods in some cases, due to small differences in distance estimates between the sphere and spheroid models. In practice, the trade-off between speed and accuracy should always be considered when choosing the right distance-based calculation given the application requirements, dataset resolution and dataset size. With small amounts of data and/or high accuracy requirements ST_Distance should be preferred; otherwise the accuracy given by <-> may be sufficient.

- What do you think will happen if we omit the ::geography keyword from both places where it appears in the above query?
- Check your answer using the CARTO SQL API.

11.4 Adding nearest points to map

We are now ready to take the SQL spatial query from Section 11.3.2 and incorporate it into the mapClick function, so that the nearest 25 plants are displayed along with the red

---

[8]Note that this query does not use data from any table, because it calculates distance between two points created as part of the query itself.

[9]https://en.wikipedia.org/wiki/Longitude#Length_of_a_degree_of_longitude

marker. The key here is that we are going to make the spatial query *dynamic*, each time replacing the proximity search according to the location clicked by the user on the web map. Unlike in Section 11.3.2, where the longitude and latitude were hard-coded into the SQL query ([34.810696, 31.895923]), we are going to generate the SQL query by pasting user-clicked coordinates into the SQL query "template". Specifically, we will use longitude and latitude returned from our map click event, `e.latlng` (Section 11.2.1).

We proceed, modifying `example-11-02.html`. First, we add two more variable definitions: a layer group for the nearest plant markers (`plantLocations`), and the URL prefix for querying the CARTO SQL API (`url`). Along with the the layer group for the clicked location (`myLocation`), which we already defined in the previous example (Section 11.2.1), the variable definition part in our `<script>` is now composed of the following three expressions:

```
var myLocation = L.layerGroup().addTo(map);
var plantLocations = L.layerGroup().addTo(map);
var url = "https://michaeldorman.carto.com/api/v2/sql?format=GeoJSON&q=";
```

The remaining code changes are all inside our `mapClick` function. In its new version, the function will not only display the clicked location, but also the locations of 25 nearest observations of rare plants. Recall that the previous version of the `mapClick` function in `example-11-02.html` (Section 11.2.1) merely cleared the old marker and added a new one according to the `e.latlng` object:

```
function mapClick(e) {
    myLocation.clearLayers();
    L.marker(e.latlng, {icon: redIcon}).addTo(myLocation);
}
```

In the new version of the function, we add four new expressions. First, we define a new local variable `clickCoords` capturing the contents of `e.latlng` (Section 6.9) for that particular event. This is just a convenience, for typing `clickCoords` instead of `e.latlng` in the various places where we will use the clicked coordinates in the new function code body:

```
var clickCoords = e.latlng;
```

Second, we make sure that the nearest plant observations are cleared between consecutive clicks, just like the red marker is. The following expression takes care of clearing any previously loaded `plantLocations` contents:

```
plantLocations.clearLayers();
```

In the third expression, we dynamically compose the SQL query to get 25 nearest records from the `plants` table, considering the clicked location. We basically replace the hard-coded lon/lat coordinates from the specific SQL query discussed in Section 11.3.2 with the `lng` and `lat` properties of the `clickCoords` object. The result, named `sqlQueryClosest`, is the specific SQL query string needed to obtain the 25 nearest plant observations from our currently clicked location. This is conceptually similar to the way that we dynamically constructed an SQL query to select the observations of a particular species (Section 10.5.3):

```
var sqlQueryClosest =
    "SELECT name_lat, the_geom FROM plants" +
    "ORDER BY the_geom::geography <-> ST_SetSRID(ST_MakePoint(" +
    clickCoords.lng + "," + clickCoords.lat +
    "), 4326)::geography LIMIT 25";
```

Fourth, we use the **sqlQueryClosest** string to request the nearest 25 observations from CARTO, and add them to the **plantLocations** layer. The popup, in this case, contains the Latin species name (**name_lat**) displayed in italics:

```
$.getJSON(url + sqlQueryClosest, function(data) {
    L.geoJSON(data, {
        onEachFeature: function(feature, layer) {
            layer.bindPopup("<i>" + feature.properties.name_lat + "</i>");
        }
    }).addTo(plantLocations);
});
```

Here is the complete code for the new version of the **mapClick** function:

```
function mapClick(e) {

    // Get clicked coordinates
    var clickCoords = e.latlng;

    // Clear map
    myLocation.clearLayers();
    plantLocations.clearLayers();

    // Add location marker
    L.marker(clickCoords, {icon: redIcon}).addTo(myLocation);

    // Set SQL query
    var sqlQueryClosest =
        "SELECT name_lat, the_geom FROM plants" +
        "ORDER BY the_geom::geography <-> ST_SetSRID(ST_MakePoint(" +
        clickCoords.lng + "," + clickCoords.lat +
        "), 4326)::geography LIMIT 25";

    // Get GeoJSON & add to map
    $.getJSON(url + sqlQueryClosest, function(data) {
        L.geoJSON(data, {
            onEachFeature: function(feature, layer) {
                layer.bindPopup(
                    "<i>" + feature.properties.name_lat + "</i>"
                );
            }
        }).addTo(plantLocations);
```

```
    });

}
```

The resulting map `example-11-03.html` is shown in Figure 11.7. The red marker denotes the clicked location, while the blue markers denote the 25 nearest plant observations loaded from the database.

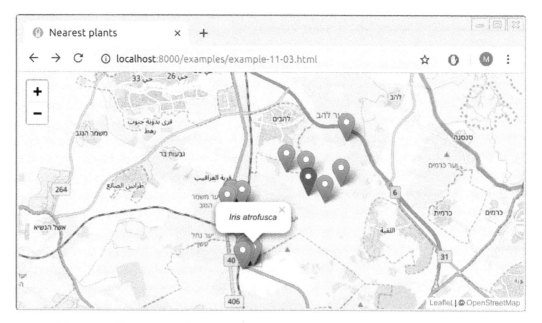

FIGURE 11.7: Screenshot of `example-11-03.html`

11.5 Drawing line connectors

To highlight the direction and distance from the clicked location to each of the nearest plants, the final version `example-11-04.html` adds code for drawing **line segments** between the focal point and each observation (Figure 11.8). This is a common visualization technique, also known as a "spider diagram" or "desire lines" (e.g., in ArcGIS documentation[10]). Line segments are frequently used to highlight the association between a focal point or points, and their associated (usually nearest) surrounding points from a different layer. For example, the lines can be used to visualize the association between a set of business stores and the customers who are potentially affiliated with each store.

To draw each of the line segments connecting the red marker (clicked location) with one of the blue ones (plant observation), we can take the two pairs of coordinates and apply the `L.polyline` function (Section 6.6.3) on them. The principle is similar to the function we wrote

---

[10]http://desktop.arcgis.com/en/arcmap/latest/extensions/business-analyst/create-spider-diagrams-desire-lines.htm

in the exercise for Chapter 3 (Section 3.12). Suppose that the clicked location coordinates are [lon0, lat0] and the coordinates of the nearest plant observations are [lon1, lat1], [lon2, lat2], [lon3, lat3], etc. The coordinate arrays of the corresponding line segments, connecting each pair of points, are then:

- [[lon0, lat0], [lon1, lat1]] for the first segment
- [[lon0, lat0], [lon2, lat2]] for the second segment
- [[lon0, lat0], [lon3, lat3]] for the third segment
- etc.

Each of these coordinate arrays can be passed to the L.polyline function to draw the corresponding segment connecting a pair of points, much like the line segment connecting Building 72 with the Library (Figure 6.9). Note that we will actually be using reversed coordinate pairs, such as [lat0, lon0], as expected by all Leaflet functions for drawing shapes, including L.polyline (Section 6.5.5).

To implement the above procedure of drawing line segments, we first create yet another layer group named **lines** at the beginning of our script, right below the expressions where we already defined the **myLocation** and **plantLocations** layer groups:

```
var lines = L.layerGroup().addTo(map);
```

Accordingly, inside the **mapClick** function we clear all previous line segments before new ones are drawn, with the following expression clearing the **lines** layer group, right below the expressions where we clear **myLocation** and **plantLocations**:

```
lines.clearLayers();
```

Moving on, we are ready to actually draw the line segments. Inside the L.geoJSON function call, in the previous **example-11-03.html** (Section 11.4) we only binded the Latin species name popup like so:

```
L.geoJSON(data, {
    onEachFeature: function(feature, layer) {

        // Bind popup
        layer.bindPopup("<i>" + feature.properties.name_lat + "<i>");

    }
}).addTo(plantLocations);
```

Now, in **example-11-04.html**, we add three more expressions inside the **onEachFeature** option. These expressions are used to draw a line segment between the "central" point where the red marker is, which is stored in **clickCoords**, and the currently-added GeoJSON point, which is given in **feature** as part of the **onEachFeature** iteration (Section 8.5):

```
L.geoJSON(data, {
    onEachFeature: function(feature, layer) {

        // Bind popup
```

```
        layer.bindPopup("<i>" + feature.properties.name_lat + "<i>");

        // Draw line segment
        var layerCoords = feature.geometry.coordinates;
        var lineCoords = [
            [clickCoords.lat, clickCoords.lng],
            [layerCoords[1], layerCoords[0]]
        ];
        L.polyline(lineCoords, {color: "red", weight: 3, opacity: 0.75})
            .addTo(lines);

    }
}).addTo(plantLocations);
```

The new code section, right below the **//Draw line segment** comment, is composed of three expressions, which we now discuss step by step.

In the first expression, we extract the coordinates of the current plant observation, and assign them to a variable named **layerCoords**. As discussed in Section 8.5, as part of the **onEachFeature** iteration, the currently processed GeoJSON feature is accessible through the **feature** parameter. Just like we are accessing the current species name with **feature.properties.name_lat** to include it in the popup, we can also extract the point coordinates with **feature.geometry.coordinates**. The **feature.geometry.coordinates** property of a GeoJSON **"Point"** geometry is an array of the form **[lon, lat]** (Section 7.3.2.2), which we store in a variable named **layerCoords**:

```
var layerCoords = feature.geometry.coordinates;
```

In the second expression, we build the segment coordinates array by combining the coordinates of the focal point stored in **clickCoords** with the coordinates of the current plant observation stored in **layerCoords**. Again, while **layerCoords** are stored as **[lon, lat]** according to the GeoJSON specification, **L.polyline** expects **[lat, lon]** (Section 6.6.3). This is why the **layerCoords** array is reversed with **[layerCoords[1], layerCoords[0]]**. The complete segment coordinates array is assigned into a variable named **lineCoords**:

```
var lineCoords = [
    [clickCoords.lat, clickCoords.lng],
    [layerCoords[1], layerCoords[0]]
];
```

In the third expressions, **lineCoords** is passed to the **L.polyline** function to create a line layer object. The **.addTo** method is then applied, in order to add the segment to the **lines** layer group and thus actually draw it on the map:

```
L.polyline(lineCoords, {color: "red", weight: 3, opacity: 0.75})
    .addTo(lines);
```

The resulting map **example-11-04.html** is shown in Figure 11.8. Clicking on the map now displays both nearest plant observations and the connecting line segments.

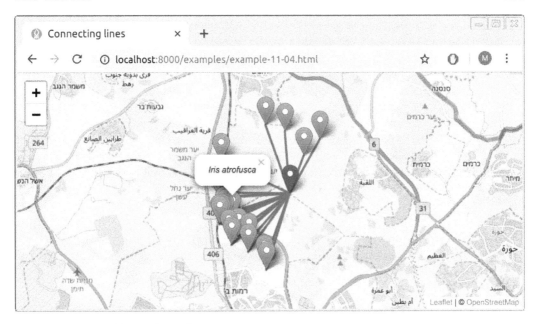

FIGURE 11.8: Screenshot of `example-11-04.html`

11.6 Exercise

- Start with `example-11-04.html` and modify the code so that the popup for each of the nearest plants also specifies its *distance* to the queried point (Figure 11.9).

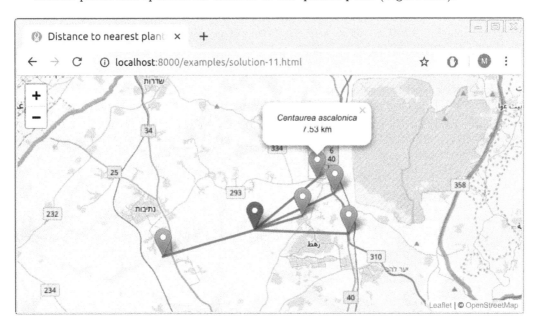

FIGURE 11.9: Screenshot of `solution-11.html`

- To get the distances, you can use the following SQL query example which returns the five nearest plants along with the distance in *kilometers* to the specific point [34.810696, 31.895923]. The distances are given in an additional column named dist_km.

```sql
SELECT name_lat,
  (
    the_geom::geography <->
    ST_SetSRID(ST_MakePoint(34.810696, 31.895923), 4326)::geography
  ) / 1000 AS dist_km,
  ST_AsText(the_geom) AS geom
FROM plants
ORDER BY dist_km
LIMIT 5;
```

- The query gives the following result. Note the dist_km column, which contains the distances in kilometers:

```
        name_lat        |      dist_km       |            geom
------------------------+--------------------+----------------------------------
 Lavandula stoechas     | 0.258166195505697  | POINT(34.808564 31.897377)
 Bunium ferulaceum      |  0.25928724896688  | POINT(34.808504 31.897328)
 Bunium ferulaceum      |  0.25928724896688  | POINT(34.808504 31.897328)
 Silene modesta         |  1.19050813583538  | POINT(34.822295 31.900125)
 Corrigiola litoralis   |  1.53676126266166  | POINT(34.825931 31.900792)
(5 rows)
```

- Here is the corresponding CARTO SQL API query you can experiment with:

```
https://michaeldorman.carto.com/api/v2/sql?format=CSV&q=
SELECT name_lat, (the_geom::geography <->
ST_SetSRID(ST_MakePoint(34.810696, 31.895923), 4326
)::geography) / 1000 AS dist_km, ST_AsText(the_geom) as geom
FROM plants ORDER BY dist_km LIMIT 5
```

- Remember that in the actual code you need to use format=GeoJSON instead of format=CSV, and the_geom instead of ST_AsText(the_geom) to get GeoJSON rather than CSV. It is also useful to round the distance values before putting them into the popup (e.g., to two decimal places, as shown in Figure 11.9). This can be done with JavaScript, applying the .toFixed method with the required number of digits on the numeric variable:

```javascript
x = 2.321342;
x.toFixed(2);  // Returns "2.32"
```

Part IV

Advanced Topics

12

Client-Side Geoprocessing

12.1 Introduction

In Chapter 11, we used spatial SQL queries to load a subset of features from a PostGIS database on a web map. In other words, we performed a geoprocessing operation—finding the n-nearest points—using SQL. Notably, the operation was performed on the **server** (the CARTO platform), and the result was returned to us via HTTP. Sometimes, it is more convenient to perform geoprocessing operations on the **client**, rather than on the server, for various reasons.

One reason to prefer client-side geoprocessing is that we may need the web page to be *instantaneously* responsive, such as a web map that is continuously updated in response to user input (Figure 12.5). This is very difficult to achieve when communicating with a server: even if the internet connection is very fast, there is usually going to be a noticeable lag due the time interval between sending the request and receiving the response. Another reason may be the relative simplicity and flexibility of using various algorithms and methods through JavaScript libraries, compared to the cost of setting up and managing a server with the same functionality.

The main disadvantage of client-side geoprocessing is that we are limited by the hardware constraints of the client. While the hardware of the server is (potentially) very powerful, and in any case it is under our control, the different clients that connect to our web page may have widely varying hardware. To make our website responsive, we should therefore limit client-side geoprocessing to relatively light-weight operations. Another constraint in client-side geoprocessing is that we are limited to JavaScript libraries and cannot rely on any other software or programming language (such as a PostGIS database).

In this chapter we will see several examples of **client-side geoprocessing**:

- Calculating Great Circle lines (Section 12.3)
- Drawing a continuously-updated TIN (Section 12.4)
- Detecting spatial clusters (Section 12.5)
- Drawing heatmaps (Section 12.6)

All of these will be done in the browser, using JavaScript code, without requiring a dedicated server. In the examples, we will use two different JavaScript libraries:

- **Turf.js** (Sections 12.2–12.5)—A general-purpose geoprocessing library, with a comprehensive set of functions for various geoprocessing tasks
- **Leaflet.heat** (Section 12.6)—A single-purpose plugin for Leaflet, dedicated to generating point-density heatmaps

12.2   Geoprocessing with Turf.js

12.2.1   Overview

Turf.js[1] is a JavaScript library for client-side geoprocessing. It includes a comprehensive set of functions, covering a wide range of geoprocessing tasks. The Turf.js library also has excellent documentation[2], including a small visual example for each of its functions.

As we will see shortly, the basic data type that Turf.js works with is GeoJSON. This is very convenient, because it means that any geoprocessing product can be immediately loaded on a Leaflet map with the `L.geoJSON` function, which we have been using in Chapters 7–11. As for the other way around, we will also see that any existing Leaflet layer can be converted back to GeoJSON using the `.toGeoJSON` method, so that it can passed to a Turf.js function for processing. Therefore, when working with Turf.js we typically have to go back and forth between Leaflet layer objects, which the Leaflet library understands, and GeoJSON objects, which the Turf.js library understands (Section 12.4.5).

In this chapter, we will use seven different functions from the Turf.js library (Table 12.1). There are dozens of other functions which we will not use; you are invited to browse through the documentation to explore what kind of other tasks can be done with Turf.js.

TABLE 12.1: Turf.js functions used in Chapter 12

Function	Description
`turf.point`	Convert point coordinates to GeoJSON of type `"Point"`
`turf.greatCircle`	Calculate a Great Circle line
`turf.randomPoint`	Calculate random points
`turf.tin`	Calculate a Triangulated Irregular Network (TIN)
`turf.clusterEach`	Apply function on each subset of GeoJSON features
`turf.convex`	Calculate a Convex Hull polygon
`turf.clustersDbscan`	Detect clusters using the DBSCAN method

12.2.2   Including the Turf.js library

To use Turf.js in our web page, we first need to include it, using a `<script>` element. As always, the JavaScript file can be included from a local copy, or from a CDN (Section 4.5.3) such as the following one:

`https://npmcdn.com/@turf/turf/turf.min.js`

We will use a local copy stored in the `js` folder on our server. Therefore, we add the following `<script>` element inside the `<head>`:

```
<script src="js/turf.js"></script>
```

---

[1] http://turfjs.org/
[2] http://turfjs.org/docs/

You can download the Turf.js file from the URL specified at the beginning of this section, or from the online version of the book (Section 0.7), and place it in a local folder if you want to include a local copy in your web page.

12.3 Great Circle line

As a first experiment with Turf.js, we will use the library to calculate a **Great Circle** line, then display it on a Leaflet web map. A Great Circle line is the shortest path between two points on the earth surface, taking the curvature of the earth into account (Figure 12.10). Through the Great Circle example, we will demonstrate the way that Turf.js functions operate on GeoJSON objects.

All functions from the Turf.js package are loaded as methods of a global object named `turf`, thus sharing the `turf.` prefix. This is much like jQuery functions start with $ (Section 4.6) and Leaflet functions start with L (Section 6.5.5).

As an example, let's open the documentation of the `turf.greatCircle` function[3]. Here is a slightly modified code of the example given for the `turf.greatCircle` function in the documentation:

```
var start = turf.point([-122, 48]);
var end = turf.point([-77, 39]);
var greatCircle = turf.greatCircle(start, end);
```

- Open the HTML document of the basic map `example-06-02.html` from Section 6.5.7.
- Add a `<script>` element in the `<head>` for including the Turf.js library.
- Open the console and execute the above three expressions, then examine the resulting objects as described below.

What does this code do? The first two expressions are using a convenience function named `turf.point`[4] to convert pairs of coordinates into a GeoJSON object of type `"Feature"`, comprising one feature of type `"Point"` (Section 7.3.2.2). The coordinates passed to `turf.point` are assumed to be of the `[lon, lat]` form, same as in GeoJSON. Typing `JSON.stringify(start)` reveals the resulting GeoJSON:

```
{
  "type": "Feature",
  "properties": {},
```

---

[3]http://turfjs.org/docs/#greatCircle
[4]http://turfjs.org/docs/#point

```
    "geometry": {
      "type": "Point",
      "coordinates": [-122, 48]
    }
  }
```

Note that the [-122, 48] coordinates passed to `turf.point` are now in the `coordinates` property of the GeoJSON. The output of typing `JSON.stringify(end)` would be identical, except for the coordinates, which will be [-77, 39] instead of [-122, 48]. (Type `JSON.stringify(end)` in the console to see for yourself.)

The third expression uses the geoprocessing function `turf.greatCircle` to calculate the Great Circle line between the two points. The result is assigned to a variable named `greatCircle`. Here is the printout of `JSON.stringify(greatCircle)`[5]:

```
{
  "type": "Feature",
  "properties": {},
  "geometry": {
    "type": "LineString",
    "coordinates": [
        [-122, 48],
        [-121.49670597260395, 48.006695584053034],
        [-120.9933027193627, 48.011192319649155],
        ...,
        [-77, 38.99999999999999]
    ]
  }
}
```

This is a GeoJSON Feature of type `"LineString"` (Section 7.3.2.2) representing a Great Circle line.

- How many [lon, lat] coordinate pairs is the `greatCircle` line composed of?
- Type the appropriate expression in the console to find out[6].

When the above code section is executed in a web page with a Leaflet map object named `map`, the following expression can be used to draw the Great Circle line we just calculated on the map. We are using `L.geoJSON` to go from a GeoJSON object to a Leaflet layer, then adding the layer on the map. Note that you need to zoom-in on the U.S. to see the line, since the line goes from Seattle to Washington, D.C.

---

[5]The line coordinates in the `greatCircle` object, except for the first three and last one, were replaced with ... to save space.

[6]The answer is 100.

```
L.geoJSON(greatCircle).addTo(map);
```

Similarly, we can draw markers at the start and end points, as follows:

```
L.geoJSON(start).addTo(map);
L.geoJSON(end).addTo(map);
```

- Type these expressions in the console, in a Leaflet web map where Turf.js was loaded and **start**, **end** and **greatCircle** were calculated.
- You should see the Great Circle line and the markers on the map.

This is the general principle of working with most functions in Turf.js, in a nutshell: reshaping our data with Turf.js convenience functions (such as **turf.point**), then passing GeoJSON objects to Turf.js geoprocessing functions (such as **turf.greatCircle**) and getting new GeoJSON objects in return. We are now ready for two extended examples demonstrating the workflow of using Turf.js with Leaflet:

- In the first example (Section 12.4), we build a web map where a continuously updated **TIN** layer is generated, while the user can drag any of the underlying points (Figure 12.5).
- In the second example (Section 12.5), we perform spatial **clustering** to detect and display distinct groups of rare species observations (Figure 12.8).

12.4 Continuously updated TIN

12.4.1 Overview

As our first extended use case of Turf.js, we are going to build a web map with a dynamic demonstration of TIN layers. The map will display a TIN layer generated from a set of randomly placed points. The user will be able to drag any of the points, observing how the TIN layer is being updated in real time (Figure 12.5). We will go through four steps to accomplish this task:

- In **example-12-01.html**, we are to going to generate **random points** and add them on a Leaflet map (Section 12.4.2).
- In **example-12-02.html**, we will generate a **TIN** layer on top of the points (Section 12.4.3).
- In **example-12-03.html**, we will learn how to make a Leaflet marker **draggable** (Section 12.4.4).

- In `example-12-04.html`, we will make all of our random points draggable and binded to an event listener that **updates** the TIN layer in response to any of the points being dragged (Section 12.4.5).

12.4.2   Generating random points

The `turf.randomPoint` function[7] from the Turf.js library can be used to generate GeoJSON with randomly placed points. When using the `turf.randomPoint` function, we need to specify:

- The **number** of points to generate
- A **bounding box**, defined using an array of the form [lon0, lat0, lon1, lat1]

Given these two arguments, the `turf.randomPoint` function randomly places the specified number of points withing the given bounding box, and returns the resulting point layer as GeoJSON. For example, in the following code section, the *first* expression defines a bounding box array named `bounds`, using the [lon0, lat0, lon1, lat1] structure that `turf.randomPoint` expects. In this case, we are using the coordinates of the bounding box of Israel (Figure 12.1). The *second* expression then generates 20 random points placed within the bounding box, and returns a GeoJSON object which we assign to a variable named `points`. Note that the bounding box array needs to be passed as the `bbox` property inside the options object in `turf.randomPoint`. If we omit the `bbox` option, the random points will be generated in the default global extent, i.e., [-180, -90, 180, 90]. The *third* expression transforms the returned GeoJSON `points` to a Leaflet layer, then adds the layer on the map:

```
var bounds = [34.26801, 29.49708, 35.90094, 33.36403];
var points = turf.randomPoint(20, {bbox: bounds});
L.geoJSON(points).addTo(map);
```

To view the points, it is convenient to set the initial map extent to the same bounding box where the points were generated. In the following expression, we initialize the Leaflet map using a view-setting method called `.fitBounds`. The `.fitBounds` method automatically detects the appropriate map center and zoom level so that the given extent fits the map bounds. The `.fitBounds` method is sometimes more convenient than `.setView`, which was introduced in Section 7.5, where the extent is set based on map center and zoom level.

Confusingly, the `.fitBounds` method from Leaflet and the `turf.randomPoint` function from Turf.js require two different forms of bounding box specifications. Specifically, like all other Leaflet functions, `.fitBounds` uses the [lat, lon] ordering instead of [lon, lat]. Also, `.fitBounds` needs the two bounding box "corners" to be separated in two internal arrays. Therefore:

- **Leaflet** needs a [[lat0, lon0], [lat1, lon1]] bounding box
- **Turf.js** needs a [lon0, lat0, lon1, lat1] bounding box

This is the reason that the `bounds` array is "rearranged" in the following call to `.fitBounds`, when initializing our Leaflet map to the same extent where the random points are[8]:

---

[7]`http://turfjs.org/docs/#randomPoint`
[8]Note that part of the attribution was replaced with [...] to save space.

```
var map = L.map("map")
    .fitBounds([[bounds[1], bounds[0]], [bounds[3], bounds[2]]]);
L.tileLayer(
    "https://{s}.tile.openstreetmap.org/{z}/{x}/{y}.png",
    {attribution: '&copy; <a href="https://www.openstreetmap.org/[...]</a>'}
).addTo(map);
```

Including the latter five expressions in our script displays 20 randomly placed point markers (`example-12-01.html`), as shown in Figure 12.1.

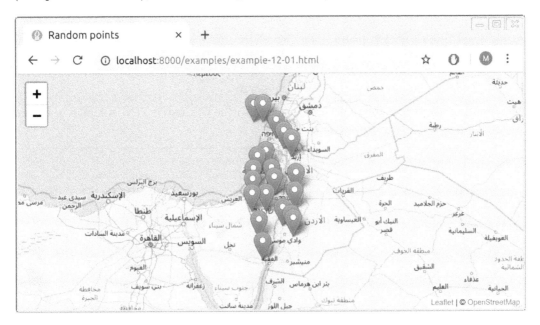

FIGURE 12.1: Screenshot of `example-12-01.html`

- Open `example-12-01.html` in the browser.
- Refresh the page several times—you should see the 20 points being randomly placed in different locations each time.

12.4.3 Adding a TIN layer

A **Triangulated Irregular Network (TIN)**[9] is a geometrical method of connecting a set of points in such a way that the entire surface is covered with triangles. TIN is frequently used in 3D modeling, as this method can be used to construct a continuous surface out of a set of 3D points. In Turf.js, the `turf.tin` function[10] can be used to calculate a TIN layer given a set of points.

---

[9]https://en.wikipedia.org/wiki/Triangulated_irregular_network
[10]http://turfjs.org/docs/#tin

Like with the `turf.greatCircle` function shown previously (Section 12.3), both input and output of `turf.tin` are GeoJSON objects. In the case of `turf.tin`, the input should be a point layer, and the output is a polygonal layer with the resulting TIN.

To draw a TIN layer based on the GeoJSON object named `points`, we can add the following expressions at the bottom of the `<script>` in `example-12-01.html`. You can also run the expressions in the console of `example-12-01.html` to see the TIN layer being added in "real-time":

```
var tin = turf.tin(points);
L.geoJSON(tin).addTo(map);
```

As shown in Section 12.4.2, `points` is a GeoJSON object representing a set of 20 random points. In the first expression, we calculate the TIN layer GeoJSON with `turf.tin(points)` and assign it to a variable named `tin`. In the second expression, `tin` is added on the map with `L.geoJSON`, the same way we added the points.

The resulting map `example-12-02.html`, now with the calculated TIN, is shown in Figure 12.2.

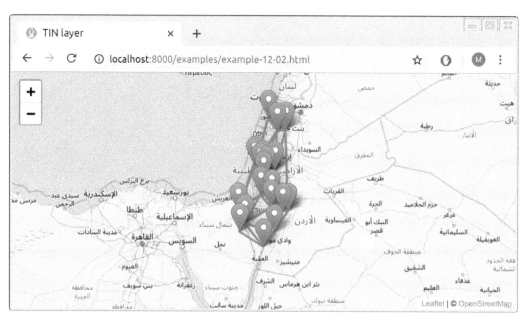

FIGURE 12.2: Screenshot of `example-12-02.html`

- Open `example-12-02.html` in the browser and refresh the page several times.
- You should see a different TIN layer each time, according to the random placement of the markers.

12.4.4 Draggable circle markers

So far, in `example-12-02.html` (Figure 12.2), we created a web map with randomly generated points and their associated TIN layer. Refreshing the web page in this example gives a nice demonstration of the TIN algorithm—each time the page is refreshed, the points are randomly rearranged and the TIN layer is updated accordingly. An even nicer demonstration, though, would be if we could *drag* any of the points wherever we want, and watch the TIN layer being updated in *real-time*. Perhaps you are already familiar with this type of behavior from **Google Maps**[11], or other web applications for routing, where the user can drag the marker denoting an origin or a destination, and the calculated route is continuously updated in response (Figure 12.3).

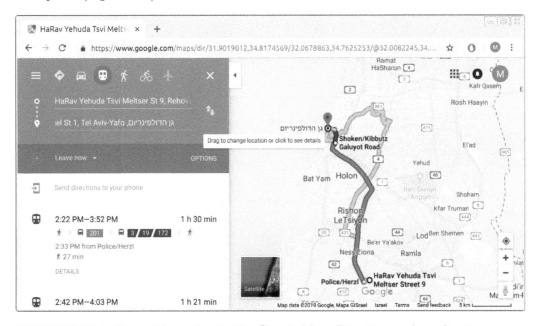

FIGURE 12.3: Draggable marker in the Google Maps Directions web application

The first thing missing to make our continuously-updated TIN is to make the 20 random markers actually *draggable*. This is exactly what we learn how to do in this section. Going back to the basic map `example-06-02.html` from Section 6.5.7, adding the following code displays a draggable marker on the map. Note that the marker is placed in a manually defined location [31.262218, 34.801472] (of the form [`lat`, `lon`]), and has a popup message. Making the marker draggable requires setting its styling option **draggable** to **true** when creating it with `L.marker`:

```
var marker =
    L.marker([31.262218, 34.801472], {draggable: true})
    .bindPopup("This marker is draggable! Move it around...")
    .addTo(map)
    .openPopup();
```

On its own, the draggable marker is not very useful, because the TIN layer will not follow it once dragged. What we would like to do is make the TIN layer respond to every change in

---
[11]https://www.google.com/maps/

marker locations, by continuously updating itself. To do that, we can add an event listener of type `"drag"` on each and every one of our 20 random markers. The event listener function we define will be executed each time any of the markers is moved around.

For now, we have just one draggable marker. To experiment with the `"drag"` event listener we can use a function that simply prints the current marker position. The latter can be accessed through a property named `._latlng` that the marker has. The `._latlng` property is an object with properties `.lat` and `.lng`, similar to the `.latlng` event object property (Sections 6.9 and 11.2.1):

```
marker.on("drag", function() {
    console.log(marker._latlng);
});
```

The resulting map `example-12-03.html`, displaying one draggable marker, is shown in Figure 12.4.

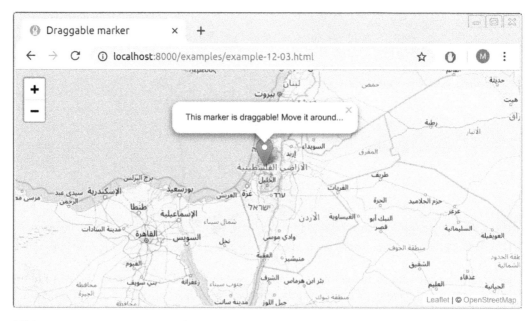

FIGURE 12.4: Screenshot of `example-12-03.html`

- Open `example-12-03.html` and try dragging the marker around.
- Open the console to see the current location being printed each time the marker is dragged.

12.4.5 Continuous updating

Now, let's combine the TIN example (Figure 12.2) with the draggable marker event listener example (Figure 12.4) to create a continuously-updated TIN. Basically, we are going to set {draggable: true} for all of the 20 randomly generated markers, then make the TIN layer *update* in response to any of those markers being dragged. The final result example-12-04 is shown in Figure 12.5.

To achieve this result, we will use a slightly modified approach for adding the points as well as the TIN layer. Instead of simply adding the points and the tin GeoJSON objects on the map right away, like we did in example-12-02.html:

```
var points = turf.randomPoint(20, {bbox: bounds});
var tin = turf.tin(points);
L.geoJSON(points).addTo(map);
L.geoJSON(tin).addTo(map);
```

we will set up two empty layer groups for the points and TIN layers, named pnt_layer and tin_layer, respectively. These layer groups will be referred to later on in the code. We are using L.layerGroup (Section 7.6.5) to create the layer groups:

```
var pnt_layer = L.layerGroup().addTo(map);
var tin_layer = L.layerGroup().addTo(map);
```

Next, we calculate the 20 random points:

```
var points = turf.randomPoint(20, {bbox: bounds});
```

This time, we do not want to convert all points to a GeoJSON Leaflet layer with the default settings of L.geoJSON. Instead, we want use L.geoJSON options to set the following:

- Make each of the markers draggable
- Add a "drag" event listener to each marker, triggering the update of the TIN layer whenever the marker is dragged

This is why instead of using L.geoJSON with the default options, like we did in example-12-02.html:

```
L.geoJSON(points).addTo(map);
```

we are now setting both the onEachFeature and pointToLayer options of L.geoJSON:

```
L.geoJSON(points, {
    onEachFeature: function(feature, layer) {
        layer.on("drag", drawTIN);
    },
    pointToLayer: function(geoJsonPoint, latlng) {
        return L.marker(latlng, {draggable: true});
```

```
    }
}).addTo(pnt_layer);
```

We are already well familiar with using the **onEachFeature** option for adding event listeners to each GeoJSON feature. For example, in Chapter 8 we used this option to add **"mouseover"** and **"mouseout"** event listeners to highlight the town polygons on mouse hover (Section 8.8.1). Here, we are adding a **"drag"** event listener. The function named **drawTIN**, which will be executed on marker drag, is yet to be defined (see below).

The other option, **pointToLayer**, was only briefly mentioned in the exercise for Chapter 8 (Sections 8.2 and 8.9). The **pointToLayer** option is used to specify the way that GeoJSON points are to be translated to Leaflet layers, in case we want to override the default **L.marker**. In this case, we just want to draw a marker with the {draggable: true} option, rather than the default marker.

Moving on, we now need to define the **drawTIN** function, which is being passed to the event listener. Each time a marker is *dragged*, the **drawTIN** function will be executed to calculate the new GeoJSON of the TIN layer, and replace the old TIN with the new one. Importantly, the new TIN layer is calculated based on the up-to-date **pnt_layer** to keep the points and the TIN *in sync*. Here is the **drawTIN** function definition:

```
function drawTIN() {
    tin_layer.clearLayers();
    points = pnt_layer.toGeoJSON();
    tin = turf.tin(points);
    tin = L.geoJSON(tin);
    tin.addTo(tin_layer);
}
```

Essentially, each time a marker is dragged, the **drawTIN** function does the following things:

- **Clears** the **tin_layer** layer group, to remove the old TIN layer
- **Calculates** the new TIN layer based on the current points in **pnt_layer**
- **Adds** the new TIN layer to the **tin_layer** layer group, thus drawing it on the map

These three things are accomplished in the 1st, 3rd, and 5th expressions in the **drawTIN** code body, respectively. What are the 2nd and 4th expressions for, then? The latter are necessary because of the above-mentioned fact (Section 12.3) that Turf.js functions only accept GeoJSON, and thus cannot directly operate on Leaflet layers. Therefore, we need to translate the Leaflet marker layer to GeoJSON in the 2nd expression, then translate the TIN GeoJSON back to a Leaflet layer in the 4th expression. The following code section repeats the internal code of **drawTIN**, this time with comments specifying the purpose of each expression:

```
tin_layer.clearLayers();         // (1) Clear old TIN
points = pnt_layer.toGeoJSON();  // (2) Layer -> GeoJSON
tin = turf.tin(points);          // (3) Calculate new TIN
tin = L.geoJSON(tin);            // (4) GeoJSON -> Layer
tin.addTo(tin_layer);            // (5) Display new TIN
```

Finally, we need to execute the `drawTIN` function one time outside of the event listener. That way, the *initial* TIN layer will be displayed on page load[12], even if the user has not dragged any of the markers yet. Without this initial function call, the TIN layer will only appear after any of the markers is dragged, but not on initial page load:

```
drawTIN();
```

The resulting map `example-12-04.html` is shown in Figure 12.5. The screenshot shows how one of the points was dragged to the left, i.e., towards the west, and the TIN layer was extended accordingly.

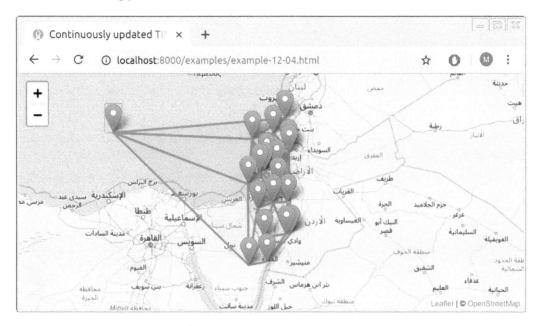

FIGURE 12.5: Screenshot of `example-12-04.html`

12.5 Clustering

12.5.1 Overview

Clustering[13] is the process of classifying a set of observations into groups, so that objects within the same group (called a cluster) are more similar to each other than to objects in other groups. With spatial clustering, similarity usually means geographical proximity, so that the aim of clustering is to group mutually proximate observations. In our second example with Turf.js, the final goal is to create a web map where clusters of nearby plant observations per species (i.e., populations) are detected using a clustering method called

---

[12]We used the same technique to display the initially selected species on the map in example `example-10-05.html` (Section 10.6).

[13]https://en.wikipedia.org/wiki/Cluster_analysis

DBSCAN (Section 12.5.4) and displayed on a web map (Figure 12.8). We will approach this task in three steps:

- In `example-12-05.html`, we will learn to automatically apply the same function on **subsets** of GeoJSON features sharing the same value of a given property, such as observations of the same species (Section 12.5.2).
- In `example-12-06.html`, we will learn to emphasize the spatial extent that a group of points occupies, by drawing **Convex Hull** polygons around the group (Section 12.5.3).
- In `example-12-07.html`, we will apply the **DBSCAN** clustering algorithm to detect spatial clusters in the observation points per species, and draw a Convex Hull polygon around each cluster (Section 12.5.4).

12.5.2   Processing sets of features

In the first step of our clustering example, we are going to load observations of four *Iris* species from the `plants` table on CARTO, which we are familiar with from Chapters 9–11. We are not doing any clustering just yet. The purpose of this example is to get familiar with the `turf.clusterEach` function[14]. The `turf.clusterEach` function is used to apply a function on each group of GeoJSON features, whereas a group is defined through common values in one of the GeoJSON properties.

The script starts with constructing the URL for the SQL API, then defining a color selection function `getColor` and styling function `setStyle` (Section 8.4), for loading and setting the color of four *Iris* species observations, respectively:

```
// Set base URL
var url = "https://michaeldorman.carto.com/api/v2/sql?format=GeoJSON&q=";

// Set SQL Query
var sqlQuery = "SELECT name_lat, the_geom " +
    "FROM plants WHERE " +
    "name_lat='Iris atrofusca' OR " +
    "name_lat='Iris atropurpurea' OR " +
    "name_lat='Iris mariae' OR " +
    "name_lat='Iris petrana'";

// Color function
function getColor(species) {
    if(species == "Iris mariae") return "yellow";
    if(species == "Iris petrana") return "brown";
    if(species == "Iris atrofusca") return "black";
    if(species == "Iris atropurpurea") return "orange";
}

// Style function
function setStyle(feature) {
    return {
        fillColor: getColor(feature.properties.name_lat),
```

---

[14]http://turfjs.org/docs/#clusterEach

```
        weight: 1,
        opacity: 1,
        color: "black",
        fillOpacity: 0.5
    };
}
```

What comes next is the important part of the script, where our GeoJSON is loaded and added on the map:

```
$.getJSON(url + sqlQuery, function(data) {
    turf.clusterEach(
        data,
        "name_lat",
        function(cluster, clusterValue, currentIndex) {
            L.geoJSON(cluster, {
                onEachFeature: function(feature, layer) {
                    layer.bindPopup("<i>" + clusterValue + "</i>");
                },
                pointToLayer: function(geoJsonPoint, latlng) {
                    return L.circleMarker(latlng);
                },
                style: setStyle
            }).addTo(map);
        }
    );
});
```

This is quite a long and complicated expression, which we will now explain in detail. First, note that the function passed to $.geoJSON is not using L.geoJSON right away, like we did before. Instead, it contains an internal function call of the turf.clusterEach function, of the following form:

```
turf.clusterEach(
    data,
    "name_lat",
    function(cluster, clusterValue, currentIndex) {
        ...
    }
);
```

The turf.clusterEach function is a convenience function from Turf.js. It is used for iterating over groups of GeoJSON features sharing the same property values. The turf.clusterEach function accepts three arguments:

- The **GeoJSON** to iterate on, in our case it is the **data** object passed from $.getJSON, i.e., the GeoJSON with the observations of four *Iris* species
- The **property** name used for grouping, in our case it is **"name_lat"**, i.e., the Latin species name

- A **function** to be applied on each set of features, with function parameters being: the current set of features (`cluster`), the current property value (`clusterValue`) and the current cluster index (`currentIndex`)

In our case, the internal function passed to `turf.clusterEach` creates and draws a GeoJSON layer, using `L.geoJSON`. The function code uses the `cluster` and `clusterValue` parameters, which refer to the current set of GeoJSON *features* sharing the same value in the `name_lat` property, and the current *value* of the `name_lat` property, respectively:

```
$.getJSON(url + sqlQuery, function(data) {
    turf.clusterEach(
        data,
        "name_lat",
        function(cluster, clusterValue, currentIndex) {
            L.geoJSON(cluster, {
                ...  // What to do with each set of features
            }).addTo(map);
        }
    );
});
```

Eventually, the code loads all observations of four *Iris* species (see SQL query above), then iterates over the four groups sharing the same species name. The iteration sequentially adds the observations of each species on the map. For example, note that when adding popups we are referring to the `clusterValue`, which refers to the `name_lat` of the current group of features in each step of the iteration:

```
layer.bindPopup("<i>" + clusterValue + "</i>");
```

You may wonder why do we need to split the GeoJSON with `turf.clusterEach` in the first place, rather than just add the entire GeoJSON with all four species at once, like we used to do in previous Chapters:

```
$.getJSON(url + sqlQuery, function(data) {
    L.geoJSON(data, {
        ...  // What to do with the entire GeoJSON
    }).addTo(map);
});
```

In this example, the latter approach is indeed preferable since it is shorter and simpler, while giving the same result. For example, it does not matter whether we apply the styling function `setStyle`, which sets the marker color, on GeoJSON subsets or on the entire object at once: the styling of a given species observation depends on its `name_lat` value either way (see Chapter 8). However, the `turf.clusterEach` approach is just a preparation for what we do in the next two sections, where it is indeed required to treat each species separately, since we will be drawing a **Convex Hull** polygon around each species (Figure 12.7) or detecting populations within each species using the **DBSCAN** clustering algorithm (Figure 12.8).

Additionally, note that when loading the GeoJSON, we are using the `pointToLayer` option. As shown in Section 12.4.5, the `pointToLayer` option controls the way that GeoJSON points are translated to Leaflet layers. Recall that in the previous example (`example-12-04.html`), we used `pointToLayer` to make our markers draggable (Figure 12.5). In the present example, the purpose of using `pointToLayer` is to override the default marker display, using *circle markers* (Sections 6.6.2 and 8.9) instead of ordinary markers. The advantage of circle markers is that they can be colored according the `clusterValue`, which refers to the `name_lat` property of the current group, so that observations of the different species are distinguished by their color:

```
...
pointToLayer: function(geoJsonPoint, latlng) {
    return L.circleMarker(latlng);
},
...
```

The resulting map `example-12-05.html` shows the observations of four *Iris* species. Each species is displayed with differently colored circle markers (Figure 12.6).

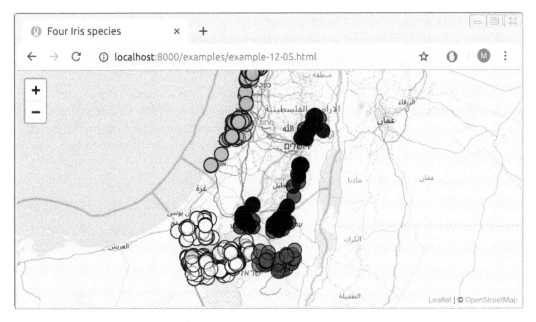

FIGURE 12.6: Screenshot of `example-12-05.html`

12.5.3   Adding a Convex Hull

A **Convex Hull**[15] is the smallest convex geometrical shape, typically a polygon, enclosing a given set of points. One of the uses of a Convex Hull in data visualization, in general, and in geographical mapping, in particular, is to highlight the geographical extent that a group of observations occupies, usually to distinguish between extents occupied by several groups. In other words, Convex Hull polygons are a useful and convenient way of highlighting the division of individual observations into *groups*.

---

[15]https://en.wikipedia.org/wiki/Convex_hull

In our second step of the clustering example (this section), we will experiment with drawing Convex Hull polygons to highlight the extent that each *Iris* species occupies in space (Figure 12.7). In the third and final step (Section 12.5.4, below), we are going to delineate internal clusters *within* each species, and draw Convex Hull polygons around each cluster within each species (Figure 12.8).

To draw a Convex Hull polygon around all observations of each species, all we need to do is add the following code within the **turf.clusterEach** iteration from **example-12-05.html**:

```
var ch = turf.convex(cluster);
ch.properties.name_lat = clusterValue;
L.geoJSON(ch, {style: setStyle}).addTo(map);
```

In the first expression we are taking the current set of points (**cluster**) and calculating their Convex Hull polygon GeoJSON using the **turf.convex** function[16]. The second expression sets the **name_lat** property of the Convex Hull according to the currently processed **clusterValue**. This step is necessary because **turf.convex** does not "carry over" the point properties from **cluster** to the resulting polygon **ch**. The **name_lat** property is needed, though, for setting the Convex Hull polygon color with **setStyle**. That way, the convex hull polygons are colored the same way as the underlying circle markers (Figure 12.7). Finally, in the third expression, we add the polygon on the map using **L.geoJSON** and **.addTo**.

Remember that **cluster** is consecutively assigned with a different set of features for each species, as part of the **turf.clusterEach** iteration. Since the code for calculating the Convex Hull and drawing it on the map is placed *inside* the iteration, a separate Convex Hull polygon is calculated per species, highlighting the separation in species distribution ranges (Figure 12.7).

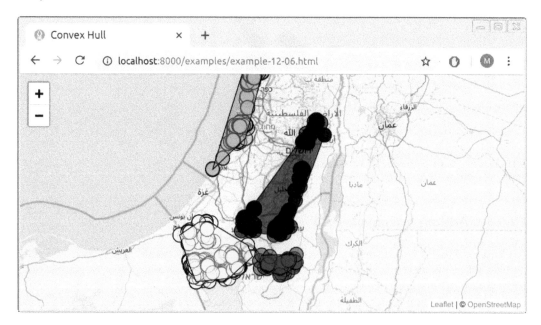

FIGURE 12.7: Screenshot of **example-12-06.html**

---

[16]http://turfjs.org/docs/#convex

12.5.4 DBSCAN clustering

In spatial clustering, we aim to detect groups of nearby observations, which are close to each other but further away from other observations. In the previous *Iris* example, we may wish to detect spatial clusters representing separate populations *within* each species. To do that, we will experiment with a clustering technique called **Density-Based Spatial Clustering of Applications with Noise** or **DBSCAN**[17].

DBSCAN is a density-based clustering method, which means that points that are closely packed together are assigned into the same cluster and given the same *ID*. The DBSCAN algorithm has two parameters, which the user needs to specify:

- ε—The maximal **distance** between points to be considered within the same cluster
- $minPts$—The **minimal number** of points required to form a cluster

In short, all groups of at least $minPts$ points, where each point is within ε or less from at least one other point in the group, are considered to be separate clusters and assigned with unique IDs. All other points are considered "noise" and are not assigned with an ID.

The `turf.clustersDbscan` function[18] is an implementation of the DBSCAN algorithm in Turf.js. The function accepts an input point layer and returns the same layer with a new property named `cluster`. The `cluster` property contains the cluster ID assigned using DBSCAN. Noise points are not assigned with an ID and thus are not given a `cluster` property. The first parameter of the `turf.clustersDbscan` function is `maxDistance` (ε), which determines what is the maximal distance between two points to be considered within the same cluster, in kilometers. In our example (see below), we will use a maximal distance of 10 kilometers. The `minPoints` ($minPts$) parameter has a default value of 3, which we will not override.

To experiment with `turf.clustersDbscan`, you can execute the following expression in the console of **example-12-01.html** (Figure 12.1). This will apply DBSCAN with a `maxDistance` value of 50 kilometers on the 20 randomly generated points:

```
turf.clustersDbscan(points, 50);
```

- Examine the result of the above expression to see which of the 20 random points were assigned to a cluster and given a `cluster` property with a numeric value (i.e., the ID).
- Try increasing or decreasing the `maxDistance` value of `50` and executing the expression again.
- As `maxDistance` gets larger, all points will tend to aggregate into the same single cluster. Conversely, as `maxDistance` gets smaller, it will be more "difficult" for clusters to be formed, which means that more and more points will be classified as noise.

Back to our *Iris* species example. When applying the DBSCAN algorithm on the observations of a given species, the non-noise observations are going to be given a `cluster` property

---

[17]https://en.wikipedia.org/wiki/DBSCAN
[18]http://turfjs.org/docs/#clustersDbscan

with a unique ID per cluster. We can then iterate over the clusters, to draw a Convex Hull polygon around each population. Practically, this can be accomplished by adding another, *internal*, `turf.clusterEach` iteration through all of the features sharing the same `cluster` attribute value *within* a given species.

To achieve the latter, we replace the three expressions we used for adding Convex Hull polygons *per species* in `example-12-06.html` (Section 12.5.3), with the following code section which adds Convex Hull polygons based on *DBSCAN clustering* per species:

```
clustered = turf.clustersDbscan(cluster, 10);
turf.clusterEach(
    clustered,
    "cluster",
    function(cluster2, clusterValue2, currentIndex2) {
        var ch = turf.convex(cluster2);
        ch.properties.name_lat = clusterValue;
        L.geoJSON(ch, {style: setStyle}).addTo(map);
    }
);
```

We leave it as an exercise for the reader to go over the code, as we are essentially combining the methods already covered in the previous two examples. The final result (`example-12-07.html`) is shown in Figure 12.8. It is now evident, for instance, how *Iris petrana* (the south-eastern species shown in brown) observations are all clustered in one place, while the distribution ranges of the other three species are divided into two or three distinct populations, separated by at least 10 kilometers from each other (Figure 12.8).

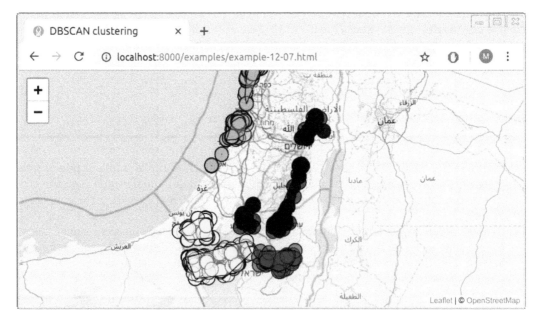

FIGURE 12.8: Screenshot of `example-12-07.html`

- Expand `example-12-07.html`, adding popups for the Convex Hull polygons to display the species name and the cluster ID when clicked.
- Modify the maximal distance parameter in `turf.clustersDbscan` (currently set to `10`) and check out how the clustering result changes. What is the reason for the console error when `maxDistance` is very small? How can we address the problem?

12.6 Heatmaps with Leaflet.heat

Turf.js is one of the most comprehensive client-side geoprocessing JavaScript libraries, but it is not the only one. The **Leaflet.heat**[19] library, for example, is a geoprocessing plugin for the Leaflet library. The term "plugin", in this context, means that Leaflet.heat is an extension of the Leaflet library and thus cannot be used without Leaflet, unlike Turf.js, which is an independent library that can be used with Leaflet or with any other library, or even on its own. Also, unlike Turf.js, which is a general-purpose geoprocessing library, Leaflet.heat does just one thing—drawing a heatmap to display the density of point data. The Leaflet.heat plugin is one of many created to extend the core functionality of Leaflet[20].

Drawing individual points can be visually overwhelming, slow, and hard to comprehend when there are too many of them. **Heatmaps** are a convenient alternative way of conveying the distribution of large amounts of point data. A heatmap is basically a way to summarize and display point density, instead of showing each and every individual point. Technically speaking, a heatmap uses two-dimensional **Kernel Density Estimation (KDE)**[21] to calculate a smooth surface representing point density.

The following example (`example-12-08.html`) draws a heatmap of all rare plant observations from the `plants` table on CARTO. Overall, there are 23,827 points in the `plants` table (Section 10.4.1). Drawing markers, or even the simpler circle markers, for such a large amount of points is usually not a good idea. First, it may cause the browser to become unresponsive due to computational intensity. Second, it is usually difficult to pick the right styling (size, opacity, etc.) for each point so that dense locations are clearly distinguished on the various map zoom levels the user may choose. This is where the density-based heatmaps become very useful. Particularly, the Leaflet.heat library automatically re-calculates the heatmap for each zoom level, allowing the user to conveniently explore point density on both large and small scales.

To use the Leaflet.heat library we first need to load it. Again, we will use a local copy:

```
<script src="js/leaflet.heat.js"></script>
```

---

[19]https://github.com/Leaflet/Leaflet.heat
[20]In Chapter 13, we are going to work with another Leaflet plugin, called **Leaflet.draw**. See the *Leaflet Plugins* section in the Leaflet documentation (https://leafletjs.com/plugins.html) for the complete list of Leaflet plugins.
[21]https://en.wikipedia.org/wiki/Kernel_density_estimation

which can be downloaded, or directly referenced, using the following URL:

`https://leaflet.github.io/Leaflet.heat/dist/leaflet-heat.js`

Next, when loading the plant observations GeoJSON we need to convert it to an array of `[lat, lon, intensity]` arrays, which is the data structure that the Leaflet.heat library expects. The `intensity` value can be used to give different *weights* to each point:

```
[
    [lat, lon, intensity],
    [lat, lon, intensity],
    [lat, lon, intensity],
    ...
]
```

There are many possible ways to go from a GeoJSON of type `"Point"` to the array structure shown above. The code section below shows one way to do it, using the `$.each` iteration method of jQuery (Section 4.12). The `$.each` iteration goes over the features of a GeoJSON object named `data`, such as the result of a CARTO SQL API request obtained with `$.getJSON`. For each of the features, the latitude and longitude are extracted into an array of the form `[lat, lon]`, along with the constant intensity value of `0.5` which gives all points equal weights when calculating density. When the iteration ends, we have an array named `locations`, containing 23,827 triplets of the form `[lat, lon, intensity]`.

```
var locations = [];
$.each(data.features, function(key, value) {
    var coords = value.geometry.coordinates;
    var location = [coords[1], coords[0], 0.5];
    locations.push(location);
});
```

Once the `locations` array is ready, we can run the `L.heatLayer` function on the array to create the heatmap layer, then add it on our map. The additional parameters define the radius of each point on the map (i.e., the smoothing kernel width) and the minimal heatmap opacity:

```
L.heatLayer(locations, {radius: 20, minOpacity: 0.5}).addTo(map);
```

The resulting map `example-12-08.html`, with the heatmap, is shown in Figure 12.9.

12.7   Exercise

- Create a web map with two markers and a Great Circle line between them, where the Great Circle line is continuously updated whenever the markers are dragged (Figure 12.10).
- Use `example-12-04.html` (Figure 12.5) for guidance on how to make the Great Circle line continuously updated.

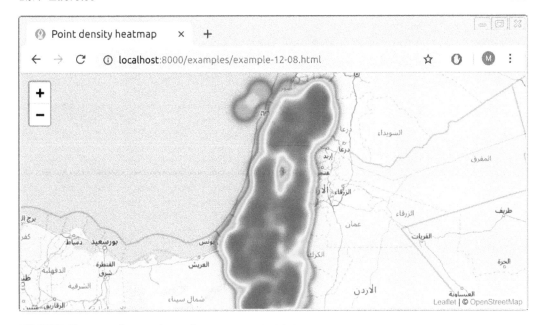

FIGURE 12.9: Screenshot of `example-12-08.html`

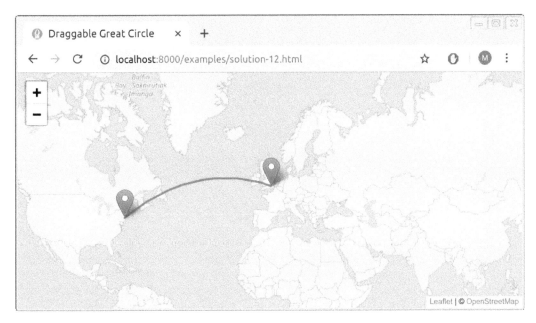

FIGURE 12.10: Screenshot of `solution-12.html`

13

Collaborative Mapping

13.1 Introduction

In this chapter, we demonstrate the construction of a special type of web maps—web maps for crowdsourcing of spatial data. Crowdsourcing web maps are used to collect, permanently store, and display spatial information contributed by numerous users. The scale of crowdsourcing can widely vary—from a web map intended for a short-term survey taken by a class or group of co-workers, up to massive collaborative projects such as OpenStreetMap where >1000 GB of information have been collected[1].

While implementing a crowdsourcing web map, we are going to learn several new concepts and techniques, including map controls for drawing ("digitizing") vector layers (Section 13.3), using a submission form (Section 13.5), and using `POST` requests for sending information to a database (Section 13.6).

13.2 Crowdsourcing

Crowdsourcing is the idea of using the power of the crowd to collect data, as an alternative to other methods, such as using data from the government or from commercial companies. The main advantages of crowdsourcing are free access to the data for all, and the ability to keep the data up-to-date. The disadvantages are unequal coverage and the risk of sabotage.

The most well-known example of crowdsourcing is **Wikipedia**[2], an encyclopedia created and edited by volunteers throughout the world, operating since 2001. The **OpenStreetMap (OSM)**[3] project, inspired by Wikipedia, was launched in 2004 to implement the idea of crowdsourcing in the field of spatial data and mapping. The aim of OSM is to create and maintain a single up-to-date digital map database of the world, through the work of volunteers, as an alternative to proprietary, out-of-date and fragmented data, predominantly used in the past. The OSM project is a big success: in many places, such as the U.S. and Western Europe, the level of detail and quality in OpenStreetMap is as good as commercial and government data sources.

From the technical point of view, crowdsourcing requires at minimum an interface where the contributors can log in and give their input, and a database where the inputs of the various contributors are being permanently stored. A crowdsourcing application needs to

---

[1] As of March 2019 (https://wiki.openstreetmap.org/wiki/Planet.osm).
[2] https://www.wikipedia.org/
[3] https://en.wikipedia.org/wiki/OpenStreetMap

be simple, accessible, and intuitive, in order to reach the broadest audience possible. A crowdsourcing application for collecting spatial data, for example, needs to be intuitive not just to GIS professionals, but also to the general public. Web applications are a good fit for crowdsourcing due to their accessibility. For example, OSM has numerous editing interfaces, known as **editors**[4]. Currently, the web-based **iD editor**[5] (Figure 13.1) is considered the default OSM editor and is responsible the largest share of OSM edits[6].

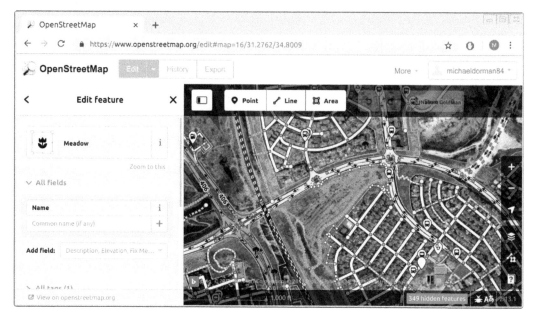

FIGURE 13.1: The iD editor, a web application for editing OpenStreetMap data

In this chapter, we are going to build a simple crowdsourcing web application. Unlike the iD editor, our crowdsourcing app will be quite minimal (Figure 13.10). For example, it will *not* have an authentication system or a complex input interface. However, the app we build will demonstrate the important concepts, which are:

- Having user **input** elements for collecting spatial data (e.g., digitizing)
- Communication with a **database** for persistently storing users' input

We are going to build the crowdsourcing application in four steps:

- In `example-13-01.html`, we will learn how to add a **drawing control** for drawing shapes on a Leaflet map (Section 13.3).
- In `example-13-02.html` and `example-13-03.html`, we will learn about translating drawn shapes to **GeoJSON** (Section 13.4).
- In `example-13-04.html`, we will add a **form** where the user can enter attribute values along with the drawn shapes (Section 13.5).
- In `example-13-05.html`, we will add a mechanism for sending the drawn shapes to a CARTO **database** for persistent storage (Section 13.6).

---

[4]http://wiki.openstreetmap.org/wiki/Editors
[5]http://ideditor.com/
[6]https://wiki.openstreetmap.org/wiki/Editor_usage_stats

13.3   The drawing control

The first thing we need for our crowdsourcing app is a vector editing toolbar. Using the toolbar, contributors will be able to draw shapes on the map, to be submitted and stored in a database later on (Section 13.6). To add an editing toolbar on top of a Leaflet map we will use the **Leaflet.draw**[7] plugin.

We start with the basic map `example-06-02.html` from Section 6.5.7. First, we need to include the Leaflet.draw JavaScript and CSS files on our page. As usual, we will use local copies of the JavaScript and CSS files, placed in the `js` and `css` directories:

```
<link rel="stylesheet" href="css/leaflet.draw.css">
<script src="js/leaflet.draw.js"></script>
```

To load the files from a CDN, the above local paths can be replaced with the following URLs:

```
https://cdnjs.cloudflare.com/ajax/libs/leaflet.draw/1.0.4/leaflet.draw.css
https://cdnjs.cloudflare.com/ajax/libs/leaflet.draw/1.0.4/leaflet.draw.js
```

Like we did when including the Leaflet library (Section 6.5.3), using the local file option also requires downloading several image files. These images are necessary to display the icons in the drawing control (Figure 13.2). The files can be obtained from `cdnjs.com`[8], or from the online version of the book (Section 0.7). The image files need to be placed in the `images` directory inside the `css` directory.

Next thing we need to have is an **editable layer** on our map. The shapes in the editable layer are the ones that we can actually edit, or supplement with new ones we draw, using the drawing control:

```
var drawnItems = L.featureGroup().addTo(map);
```

The editable layer, hereby named `drawnItems`, is a **Feature Group** object (Section 6.6.5). A Feature Group, in Leaflet, is similar to a Layer Group but with a few additional capabilities[9], which are beyond the scope of this book. For our purposes, both Layer Group and a Feature Group are used to combine several layers into one object, which facilitates actions such as clearing the map of all layers of a given type (Sections 7.6.5 and 10.5.3). In the present example, we are using a Feature Group only because it is required by the Leaflet.draw plugin. As shown in the above expression, a feature group is created and added to the map exactly the same way as a layer group, except that we are using the `L.featureGroup` function instead of the `L.layerGroup` function.

---

[7]https://leaflet.github.io/Leaflet.draw/docs/leaflet-draw-latest.html
[8]https://cdnjs.com/libraries/leaflet.draw
[9]https://leafletjs.com/reference-1.5.0.html#featuregroup

Initializing the drawing control itself is done as follows:

```
new L.Control.Draw({
    edit: {
        featureGroup: drawnItems
    }
}).addTo(map);
```

The `new` keyword is a way to create a new object in JavaScript, which we have not seen until now. This method is a little different from creating objects through calling a **constructor** function, which is what we did so far (e.g., using `L.map`, `L.marker`, etc., without the `new` keyword). The practical differences between the two initialization methods are beyond the scope of this book. Again, we are using the `new` initialization method only because the `Leaflet.draw` plugin requires it.

The above expression initializes the drawing control and places it in the top-left corner of the web map (Figure 13.2), similarly to the way that we initialized and added other types of controls throughout the book—a map description (Section 6.8), a legend (Section 8.6), an information box (Section 8.8.2), and a dropdown menu (Section 10.3). Upon initialization, the `L.Control.Draw` function accepts an options object where we can alter various settings to specify the structure and behavior of the drawing control. In our case, we set just one option: the fact that `drawnItems` is the Feature Group that stores all editable shapes.

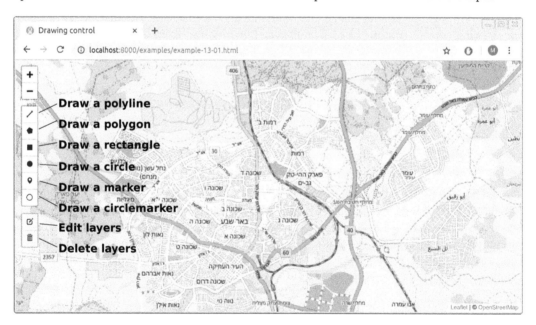

FIGURE 13.2: The Leaflet.draw control

If we added just the above two code sections, we would see the drawing control and could use it to draw shapes. However, once any of the shapes is finished it immediately disappears from the map. What we would usually like to have, instead, is that the drawn shapes persist on the map. That way, we can examine our creation: whether the drawing looks right, or whether we need to edit it further on. The key for setting up this behavior is the fact that interacting with the drawing control fires custom events[10], which we can listen to and respond to (Section 6.9).

For example, creating any new shape with the drawing control fires the `draw:created` event in the browser. Furthermore, the *event object* (Section 4.11) for the `draw:created` event contains a property named `layer`, which contains the newly drawn shape. Whenever the event fires, we can "capture" the newly drawn `e.layer` and add it to the `drawnItems` group, so that it persists on the map. This automatically makes the layer *editable*, since we set `drawnItems` as an editable group when initializing the drawing control (see above). The event listener for adding drawn shapes to `drawnItems` can be defined as follows:

```
map.on("draw:created", function(e) {
    e.layer.addTo(drawnItems);
});
```

Together, the last three code sections initialize a drawing control inside our basic map, and set an editable layer where the drawn items are "collected" and displayed on the map. The result (`example-13-01.html`) is shown in Figure 13.3.

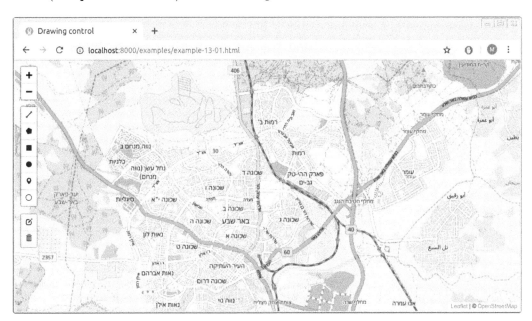

FIGURE 13.3: Screenshot of `example-13-01.html`

- Open `example-13-01.html` in the browser.
- Try drawing each of the six shape types: a line, a polygon, a rectangle, a circle, a marker, and a circle marker.
- Click on the **edit layers** button (Figure 13.2) to edit the shapes that you have drawn.
- Click on the **delete** button (Figure 13.2) to delete shapes.

---

[10]https://leaflet.github.io/Leaflet.draw/docs/leaflet-draw-latest.html#l-draw-event

- Inspect the `drawnItems` object, which contains the currently drawn shapes, in the JavaScript console.
- Run the expression `drawnItems.toGeoJSON()` to inspect the drawn shapes as GeoJSON.

13.4   Working with drawn items

13.4.1   Printing drawn items GeoJSON

In the previous example (`example-13-01.html`), each time a new shape was drawn it was simply added to our `drawnItems` Feature Group. As a result, the shape was displayed on the map, and we could subsequently edit or delete it, as well as draw any number of additional shapes. Our next step is to see how we can do something more with the drawn shapes. We will learn how to access the geometry of the drawn shapes, convert it to a GeoJSON string, and do something with it, such as to print it in the console.

In `example-13-01.html`, the currently drawn shapes were contained in the `drawnItems` Feature Group. To print the GeoJSON of the drawn shapes, the only place in the code we need to modify is the `draw:created` event listener. Instead of the original event listener definition (Section 13.3 above), we can use the following, expanded one:

```
map.on("draw:created", function(e) {
    e.layer.addTo(drawnItems);
    drawnItems.eachLayer(function(layer) {
        var geojson = JSON.stringify(layer.toGeoJSON().geometry);
        console.log(geojson);
    });
});
```

The novel part in this version is the second internal expression, starting with `drawnItems.eachLayer`. The `.eachLayer` method of a Feature Group (which a Layer Group also has), is a convenient method for doing something with each layer in the group. The `.eachLayer` method takes a function with one parameter (`layer`). The function is then applied to each of the layers in the object. This type of iteration should already be familiar; for example, the `$.each` function from jQuery (Section 4.4) or the `turf.clusterEach` function from Turf.js (Section 12.5.2) are conceptually similar iteration methods. Conveniently, each layer in `drawnItems` is by definition a separately drawn shape[11], since each layer was added when we finished drawing a shape, and the `"draw:created"` event was fired.

---

[11]The reason for using an iteration, rather than converting the entire Feature Group into a single GeoJSON, is that each geometry needs to be submitted as a separate entry in the database. This will become clear in our final example (Section 13.6.2).

Inside the function, which is applied on each layer, there are two expressions:

```
var geojson = JSON.stringify(layer.toGeoJSON().geometry);
console.log(geojson);
```

These two expression actually do a lot:

- Converting the current **layer** to GeoJSON with the **.toGeoJSON** method (Section 12.4.5)
- Selecting just the **"geometry"** property of the GeoJSON, using **.geometry** (Section 7.3.2)
- Applying the **JSON.stringify** function to convert the GeoJSON geometry object to a string (Section 3.11)
- Printing the string with **console.log**

The resulting map (**example-13-02.html**) is shown in Figure 13.4. The screenshot shows the GeoJSON string printed in the console after one shape (a rectangle) has been drawn on the map.

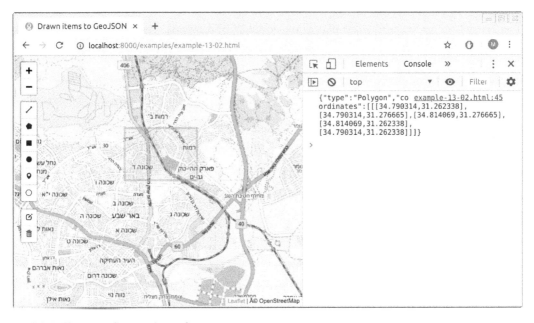

FIGURE 13.4: Screenshot of **example-13-02.html**

- Open **example-13-02.html** in the browser.
- Draw several shapes and inspect the output printed in the console.

13.4.2   Expanded GeoJSON viewer

As another example of doing something with the drawn shapes, we can now expand the GeoJSON viewer `example-07-02.html` (Figure 7.5) from Section 7.6. In that example, we had a text area where the user could type GeoJSON strings. When clicking "submit", the typed GeoJSON string was turned into a Leaflet layer and displayed on the map. Now that we know how to work with the drawing control, we can implement the other direction too: letting the user draw new shapes on the map and making the GeoJSON strings of those drawn shapes appear in the text area (Figure 13.5).

To make things simple, we will decide the user can only draw one geometry at a time, and cannot edit or delete it after creation. To do that, we define the drawing control a little differently, with editing disabled:

```
new L.Control.Draw({
    edit: false
}).addTo(map);
```

The event listener we need to display the GeoJSON of drawn shapes inside the text area is as follows:

```
function showText(e) {
    layers.clearLayers();
    var layer = e.layer;
    layer.addTo(layers);
    var geojson = JSON.stringify(layer.toGeoJSON().geometry, null, 4);
    $("#geojsontext").val(geojson);
}
map.on("draw:created", showText);
```

The above code section binds a `"draw:created"` event listener to the map, implying that the `showText` function will be executed each time a new shape is drawn. Let us review the code body of the `showText` function, expression by expression:

- `layers.clearLayers()` clears the map of all previously drawn shapes.
- `var layer = e.layer` captures the last drawn layer, using the `.layer` property of the event object, and assigns it to a variable named `layer`.
- `layer.addTo(layers)` adds the last drawn layer to the `layers` layer group, which stores the layers to be displayed on the map.
- `JSON.stringify(layer.toGeoJSON().geometry, null, 4)` extracts the GeoJSON text string of the last drawn geometry. The additional (..., `null, 4`) parameters in `JSON.stringify` make the output string indented and split into several lines (Section 3.11.2).
- `$("#geojsontext").val(geojson)` replaces the current contents of the text area with the GeoJSON string of the last drawn shape.

The new version of the GeoJSON viewer (`example-13-03.html`) is shown in Figure 13.5. The screenshot shows a line, created using the drawing control, and the corresponding GeoJSON that automatically appears in the text area.

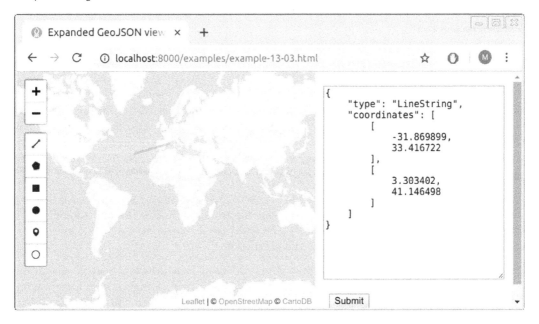

FIGURE 13.5: Screenshot of `example-13-03.html`

When experimenting with the translation of different types of shapes to GeoJSON in `example-13-03.html`, you may note two peculiar things:

- **Markers**, **circles**, and **circle markers** are all converted to `"Point"` geometries, which means that the distinction between these three layer types is not recorded in the GeoJSON.
- A **circle** is converted to a `"Point"` geometry without the circle radius being recorded in the GeoJSON.

In other words, drawn markers, circles, and circle markers all become `"Point"` geometries when converted to GeoJSON, where their original shape type and radius are not being kept.

Another thing to note is that the Leaflet.draw plugin only supports single-part geometries. In other words, using Leaflet.draw you can only create `"Point"`, `"LineString"`, and `"Polygon"` geometries. The other four GeoJSON geometry types, including the multi-part geometries `"MultiPoint"`, `"MultiLineString"`, and `"MultiPolygon"`, as well as `"GeometryCollection"`, cannot be created[12].

For simplicity and consistency with the GeoJSON format, which we are going to use to record drawn shapes and send them to the database, it therefore makes sense to only allow drawing of three shape types:

- Markers, which will be converted to `"Point"` GeoJSON
- Lines, which will be converted to `"LineString"` GeoJSON
- Polygons, which will be converted to `"Polygon"` GeoJSON

Therefore, in the next two examples (Sections 13.5–13.6), we will restrict the drawing control to just these three shape types.

---

[12]If necessary, the drawn shapes can always be combined into their *multi-* counterparts programmatically, using JavaScript code. For example, the `turf.combine` function from Turf.js (Chapter 12) can be used to combine geometries.

13.5   Submission form

In addition to the geometry, a crowdsourcing web application usually collects non-spatial attribute data too. For example, when contributing to OpenStreetMap, you typically draw ("digitize") a geometry (a point, line, or polygon), then type in its non-spatial attributes, or **tags**[13] in OSM terminology. For example, when adding a new building to the OSM data you may include the `building=yes` tag, specify building height with the `building:height` tag, and so on.

In our next example (`example-13-04.html`), we will include a simple **form** (Section 1.6.12), where the user enters a description and the contributor's name. The description and name entered by the user are going to be sent to the database as non-spatial attributes, together with the GeoJSON representing the drawn geometry.

Like in the previous examples, the first thing we need is a drawing control definition. This time we will disable the buttons for adding circles, circle markers and rectangles. We keep the buttons for adding markers, lines, and polygons, which can be translated to `"Point"`, `"LineString"`, and `"Polygon"` GeoJSON geometries, respectively, without loss of information (Section 13.4.2):

```
new L.Control.Draw({
    draw : {
        polygon : true,
        polyline : true,
        rectangle : false,    // Rectangles disabled
        circle : false,       // Circles disabled
        circlemarker : false, // Circle markers disabled
        marker: true
    },
    edit : {
        featureGroup: drawnItems
    }
}).addTo(map);
```

Our submission form will reside in a popup, conveniently appearing on top of the drawn shape once it is finished. The form contents will be defined with the following HTML code:

```
<form>
    Description:<br><input type="text" id="input_desc"><br>
    Name:<br><input type="text" id="input_name"><br>
    <input type="button" value="Submit" id="submit">
</form>
```

You can see what the form looks like in Figure 13.6.

---

[13]https://wiki.openstreetmap.org/wiki/Tags

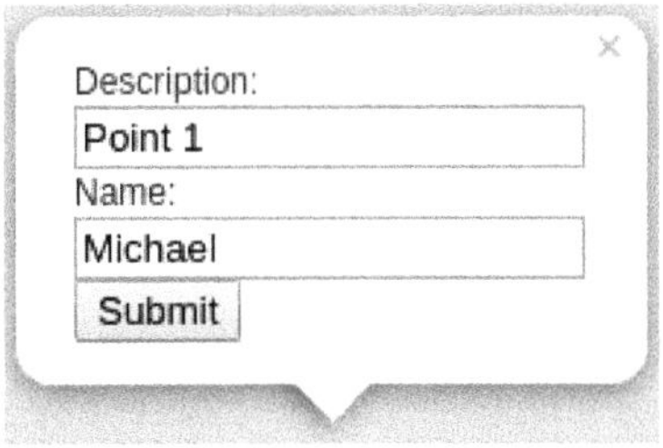

FIGURE 13.6: A form for collecting non-spatial attributes inside a Leaflet popup

Based on the above HTML code for creating the form, we can write a function called `createFormPopup` that binds and opens a popup with an editable form on the `drawnItems` feature group:

```
function createFormPopup() {
    var popupContent =
        '<form>' +
        'Description:<br><input type="text" id="input_desc"><br>' +
        'Name:<br><input type="text" id="input_name"><br>' +
        '<input type="button" value="Submit" id="submit">' +
        '</form>';
    drawnItems.bindPopup(popupContent).openPopup();
}
```

The `createFormPopup` function will be executed each time the user finishes to draw a shape, so that he/she can also enter the name and description and submit everything to the database. Therefore, we slightly modify the `"draw:created"` event listener, so that when done editing, the "submit" popup will open. Basically, we replace the `drawnItems.eachLayer` iteration inside the `"draw:created"` event listener defined in `example-13-02.html`, with a `createFormPopup` function call. That way, instead of printing the drawn shape GeoJSON in the console, an editable popup will open:

```
map.on("draw:created", function(e) {
    e.layer.addTo(drawnItems);
    createFormPopup();
});
```

What's left to be done is to take care of what happens when the user clicks the *submit* button on the form. First, we need to define an event listener for handling clicks on the `#submit` button. The event listener refers to the `setData` function, which we will define shortly. This event listener has a slightly different structure than what we used until now. The event listener is binded to the `<body>` element, i.e., the entire page contents, rather than the button itself. In addition, the `.on` method is called with an additional parameter `"#submit"`. This acts as a filter, constraining the event listener to respond only to the descendants of `<body>` which correspond to the `#submit` selector, that is, just to the button element:

```
$("body").on("click", "#submit", setData);
```

You may wonder why we cannot just we use the simpler form of an event listener which we have used until now, such as:

```
$("#submit").on("click", setData);
```

The reason is that the popup contents, including the "submit" button, are only added to the DOM *after* the user has drawn a shape and the `createFormPopup` function was executed. The last expression will therefore have no effect, since when the event listener is being binded (on page load) the popup does not exist yet. Binding the event listener to `<body>`, while restricting it to descendant element matching the `#submit` filter, solves the problem.

The event listener for clicking the "submit" button triggers the `setData` function. This is the part where we determine what to do with the drawn layer. Here is the definition of the `setData` function:

```
function setData() {

    // Get user name and description
    var enteredUsername = $("#input_name").val();
    var enteredDescription = $("#input_desc").val();

    // Print user name and description
    console.log(enteredUsername);
    console.log(enteredDescription);

    // Get and print GeoJSON for each drawn layer
    drawnItems.eachLayer(function(layer) {
        var drawing = JSON.stringify(layer.toGeoJSON().geometry);
        console.log(drawing);
    });

    // Clear drawn items layer
    drawnItems.closePopup();
    drawnItems.clearLayers();

}
```

The `setData` function collects all of the user entered information into three variables, and just prints them in the console (for now):

- `enteredUsername`—The text input from the `#input_name` field of the popup
- `enteredDescription`—The text input from the `#input_desc` field of the popup
- `drawing`—The geometry of each drawn shape, as GeoJSON

The first two inputs, `enteredUsername` and `enteredDescription`, are extracted from the text input area using the `.val` method (Section 4.7.7). The `drawing` variable is created a little differently, inside an `.eachLayer` iteration (Section 13.4.1). Using an iteration is essential, since the `drawnItems` feature group may contain more than one layer in case the user has drawn more than one shape. In such case, each layer is converted into a separate GeoJSON and separately printed in the console.

In the resulting web map `example-13-04.html` (Figure 13.7), the user can draw one or more shapes with the drawing control. Each time a shape is drawn, a popup with the *name* and *description* text inputs is opened. Once the user decides to click "submit", the name and description, as well as each GeoJSON for the drawn shapes, are printed in the console. Finally, the popup is closed, and the layer is removed from the map using the `.closePopup` and `.clearLayers` methods in the last two expressions.

One minor inconvenience is that the submission popup remains open while entering the "edit" or "delete" modes. This is counter-intuitive, as we should not submit the drawing while still editing it and making changes. The following code binds a few additional event listeners to close the popup when the user enters the "edit" or "delete" modes, and re-open it when done editing or deleting. This is accomplished using `"draw:editstart"`, `"draw:deletestart"` event listeners combined with the `.openPopup` method on the one hand, and the `"draw:editstop"` and `"draw:deletestop"` event listeners combined with the `.closePopup` methods on the other hand.

As you can see in the code section that follows, the `"draw:deletestop"` event listener is slightly more complex: its internal code contains a conditional (Section 3.10.2) for checking whether the user has deleted *all* of the drawn shapes. In the latter case, running `drawnItems.openPopup()` would cause an error, since there are no layers to open the popup on. Therefore, a conditional is first being evaluated to verify that at least one layer remains when the user is done deleting. If there is at least one layer—the editable popup will open. If there are no layers left—nothing will happen; the user will see an empty map where he/she can draw new shapes once more.

```
map.on("draw:editstart", function(e) {
    drawnItems.closePopup();
});
map.on("draw:deletestart", function(e) {
    drawnItems.closePopup();
});
map.on("draw:editstop", function(e) {
    drawnItems.openPopup();
});
map.on("draw:deletestop", function(e) {
    if(drawnItems.getLayers().length > 0) {
        drawnItems.openPopup();
    }
});
```

The result (`example-13-04.html`) is shown in Figure 13.7.

- Open `example-13-04.html` in the browser.
- Draw several shapes, then fill-in the name and description and press the "submit" button on the popup.
- Inspect the output printed in the console.
- Try editing or deleting some of the shapes, then pressing "submit" once more; the output printed in the console should always reflect the up-to-date drawn shapes as shown on the map.

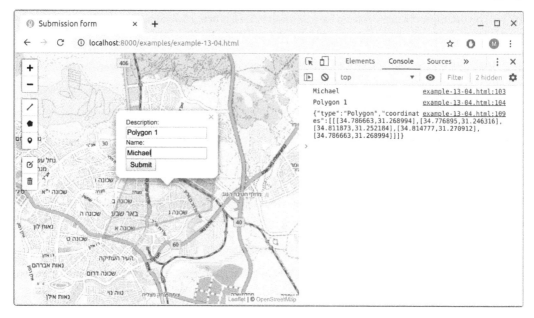

FIGURE 13.7: Screenshot of `example-13-04.html`

13.6   Sending features to the database

13.6.1   Setting database permissions

In the last example (`example-13-04.html`), the drawn layers were not really sent anywhere, just printed in the console. Other than an input interface, to have a functional crowdsourcing application we also need a permanent storage location and a mechanism for writing user input into that location. Through the rest of this chapter, we will see how the drawn shapes can be sent to a CARTO database, making the user input persistently stored, thus finalizing our crowdsourcing app.

Before we begin writing any JavaScript code for sending the data, we need to have a permanent storage location to collect the data. When using a relational database as permanent storage, what we need is an (empty) table in our database, having the corresponding columns and data types according to the data we intend to collect. On CARTO, we can create a new table (Figure 13.8) with the **New Dataset** button which we already used to upload an existing layer, such as `plants.geojson` (Section 9.7.3). This time, instead of uploading a GeoJSON file we need to choose **CREATE EMPTY DATASET**. The new table is empty, and conveniently it comes with the default `the_geom` (geometry), `description` (string), and `name` (string) columns, which is exactly what we need. In case we needed a different set of columns, we could always add or remove columns, or change column data types, using the CARTO interface. We will call the new table `beer_sheva`.

In case we are working with an SQL database through the command line, rather than the CARTO web interface, we could also create the `beer_sheva` table using the equivalent SQL command:

FIGURE 13.8: Creating an empty table on CARTO

```
CREATE TABLE beer_sheva(
  the_geom geometry,
  description text,
  name text
);
```

Now, let us suppose that the crowdsourcing web map is ready and the user has drawn a point, which we decoded into the following GeoJSON string with JSON.stringify, like we do in example-13-04.html:

```
{"type":"Point","coordinates":[34.838848,31.296301]}
```

Subsequently, let us also suppose that the user has filled the values "Point 1" and "Michael" into the description and name fields of the popup form (Figure 13.6), respectively. How can we actually insert these data into the newly created CARTO table? We can use the SQL INSERT INTO and VALUES keywords for inserting new data, as shown in the following SQL query example:

```
INSERT INTO beer_sheva (the_geom, description, name) VALUES (
  ST_SetSRID(
    ST_GeomFromGeoJSON(
      '{"type":"Point","coordinates":[34.838848,31.296301]}'
    ),
    4326
  ),
  'Point 1',
```

```
  'Michael'
);
```

The query looks quite long and complex, but note the high-level structure used to specify the column names and values to insert:

```
INSERT INTO beer_sheva(..., ..., ...) VALUES (..., ..., ...);
```

The first three ... symbols are replaced with the column names where the values go into. The last three ... symbols are replaced with the values themselves. Note that the order of column names needs to match the order of values, so that the right value will be inserted into the right column. In the present example, the ordering of the first triplet (the column names `the_geom`, `description`, and `name`) matches the order of the second triplet after the `VALUES` keyword (the geometry, `'Point 1'`, and `'Michael'`).

To create the geometry value which goes into the `the_geom` column, the query makes use of the `ST_GeomFromGeoJSON` function to convert from GeoJSON into WKB. This is exactly the opposite of decoding the geometry column into GeoJSON text with `ST_AsGeoJSON` (Section 9.6.3). The `ST_SetSRID` function specifies that our GeoJSON coordinates are in lon/lat, i.e., in the WGS84 coordinate reference system which we work with throughout the book (Section 11.3), specified with the EPSG code `4326`.

The corresponding CARTO SQL API query is given below:

```
https://michaeldorman.carto.com/api/v2/sql?q=
INSERT INTO beer_sheva (the_geom, description, name)
VALUES (ST_SetSRID(ST_GeomFromGeoJSON(
'{"type":"Point","coordinates":[34.838848,31.296301]}'
),4326),'Point 1','Michael')
```

Alas, the above API call will not work on a newly created table. Instead, an error message such as the following one will be returned:

```
{"error":["permission denied for relation beer_sheva"]}
```

This error message, as the phrase `"permission denied"` suggests, concerns the issue of database **permissions**, which we have not really considered yet. Any database, in fact, is associated with one or more database *users*, with each user having his/her own *password* and associated with a set of *privileges*, i.e., rules for what the user can and cannot do in the database. For example, an **administrator** may have the maximal set of privileges, meaning that he/she can do anything in the database: reading and writing into tables, creating new tables, deleting existing tables, adding or removing other users, granting or revoking privileges to other users, and so on. On the other hand, a **read-only user** may have limited "read-only" privileges, so that they can only consume content from the database but cannot make any changes in the tables, or in their own or other users' privileges.

The way we accessed our CARTO database with the CARTO SQL API, in fact, implied a database connection with the default user named `publicuser`, which is automatically created by CARTO when setting up our account. The `publicuser` user has *read* permissions on all tables in our database, which is why we could execute any of the SQL queries starting with `SELECT` throughout Chapters 9–12. The `publicuser`, however, by default does not have write permissions. This is why the above `INSERT` query failed with a `"permission denied"` error.

In addition to `publicuser`, CARTO defines an **API Key**[14] user, who has all possible privileges on the tables in the database: read, write, update, create, delete, and so on. To use the CARTO SQL API with the "API Key" user, we need to supply an additional `api_key` parameter in our query, as in:

```
https://michaeldorman.carto.com/api/v2/sql?q=
INSERT INTO beer_sheva (the_geom, description, name)
VALUES (ST_SetSRID(ST_GeomFromGeoJSON(
'{"type":"Point","coordinates":[34.838848,31.296301]}'
),4326),'Point 1','Michael')&
api_key=fb85********************************
```

The `api_key` is a long string, which acts like a password. You can get the API Key from your account settings panel in the CARTO web interface.

So, there are actually two possible solutions to the `"permission denied"` problem when trying to insert a new record into the `beer_sheva` table:

- We can connect to the database with the **API Key** user, who has maximal privileges and therefore can execute INSERT queries on the `beer_sheva` table.
- We can **grant** the `publicuser` user a new privilege, for executing INSERT queries on table `beer_sheva`.

The first option may seem the most convenient, since the only thing we need to do is locate our API Key string in the CARTO interface and attach it in the SQL API query URL, as shown above. However, there is a serious security issue we need to consider when using this approach. If we include the API Key in our JavaScript code, in principle anyone looking into the source code of our page will be able to copy the API Key, and use it to make any kind of SQL query on our account, having maximal privileges. For example, if anyone wanted to, he/she could even permanently delete any table in our account using `DROP TABLE` command. Exposing the API Key in *client-side* scripts is therefore a serious security risk. The API Key is really intended only for *server-side* scripts, whose source code cannot be accessed by the web page users. For instance, the server-side script may accept requests with a password the user entered; if the password is valid the server can make a query to the CARTO SQL API and send back the result, otherwise the query will be rejected. This approach requires setting up a dynamic server (Section 5.4.3), which means that, to use it securely, the API Key solution is not so simple after all. In this book, we concentrate on client-side solutions, so we will not elaborate further on the API Key approach.

For a simple crowdsourcing app, intended for a trusted audience, the second option of granting INSERT privileges to `publicuser` is a simple and effective solution. In a way, this makes our database exposed: anyone who enters our web page will be able to insert new records into the `beer_sheva` table. On the other hand, the worst-case scenario is just that our table will be filled with many unnecessary records. The only privilege we will grant is INSERT, which means that `publicuser` cannot delete any previously entered records or modify the table in any other way. Moreover, when the URL for our page is shared with a trusted audience, such as among students taking a survey in a class, the chances of someone taking the trouble of finding our page and intentionally sabotaging our database by filling it with a large amount of fake records is very small. Thus, in small-scale use cases, the effort of making a dynamic server with an authentication system may be superfluous, since the simple solution presented below is sufficient.

---

[14]`https://carto.com/developers/sql-api/guides/authentication/`

To grant the permission for making `INSERT` queries on the `beer_sheva` table, the following SQL query needs to be executed. The `publicuser` obviously does not have the permission to grant himself with additional privileges. Therefore the query needs to be executed inside the **SQL editor** on the CARTO web interface, which implies full privileges (Figure 13.9), or using the SQL API with the the API Key.

```
GRANT INSERT (the_geom, description, name)
  ON beer_sheva
  TO publicuser;
```

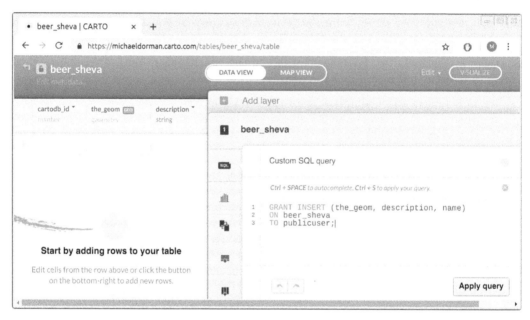

FIGURE 13.9: Granting `INSERT` permission through the CARTO web interface

After the query for granting new permissions is executed, the above `INSERT` API call, for adding a new record into the `beer_sheva` table, should work even without the API Key. In case you still see a message such as:

`{"error":["permission denied for sequence beer_sheva_copy_cartodb_id_seq"]}`

you need to execute the following query too, either in the web interface or with the API Key[15]:

```
GRANT USAGE ON SEQUENCE beer_sheva_copy_cartodb_id_seq
  TO publicuser;
```

To disable the new permission, for example when data collection is completed and we do not want to accept any more entries into the `beer_sheva` table, we can always **revoke** the privilege granted to `publicuser` as follows:

---

[15]Replace `beer_sheva_copy_cartodb_id_seq` with the exact name that was given in your error message.

```
REVOKE INSERT ON TABLE beer_sheva
  FROM publicuser;
```

After the `beer_sheva` table is created and the `INSERT` permissions are taken care of, the database is ready and waiting to receive the crowdsourced data. Now, let us move on to defining the web-map script for processing and sending the user-edited data to the database.

13.6.2   Adding the drawing control

Like in the previous examples, first thing we need in our script is to have a drawing control definition. We use the same definition from `example-13-04.html` (Section 13.5), with circles, circle markers, and rectangles disabled[16]:

```
new L.Control.Draw({...}).addTo(map);
```

Next, in case we would like the map to display the previous shapes drawn by other users, we can set up a layer group named `cartoData` to contain them:

```
var cartoData = L.layerGroup().addTo(map);
```

Accordingly, we need to load the items already contained in the `beer_sheva` table from previous editing sessions, using the following code section:

```
var url = "https://michaeldorman.carto.com/api/v2/sql?format=GeoJSON&q=";
var sqlQuery = "SELECT the_geom, description, name FROM beer_sheva";
function addPopup(feature, layer) {
    layer.bindPopup(
        feature.properties.description +
        "<br>Submitted by " + feature.properties.name
    );
}
$.getJSON(url + sqlQuery, function(data) {
    L.geoJSON(data, {onEachFeature: addPopup}).addTo(cartoData);
});
```

Note that the popup for each of the loaded features displays the **name** and the **description** properties, which were entered in previous sessions when submitting drawn shapes and saved in the `beer_sheva` table.

Other than loading previously stored shapes using the above code, the major change compared to `example-13-04.html` is in the `setData` function, which is responsible for saving the drawn shapes whenever the "submit" button is clicked. The new version is quite longer, since instead of just printing the data in the console it now sends the data for permanent storage in the CARTO database:

---

[16]The `L.Control.Draw` options were replaced with ... to save space, see Section 13.5 for the complete code.

```
function setData() {

    // Get user name and description
    var enteredUsername = $("#input_name").val();
    var enteredDescription = $("#input_desc").val();

    // Create SQL expression to insert layer & send data
    drawnItems.eachLayer(function(layer) {
        var drawing = JSON.stringify(layer.toGeoJSON().geometry);
        var sql =
            "INSERT INTO beer_sheva (the_geom, description, name) " +
            "VALUES (ST_SetSRID(ST_GeomFromGeoJSON('" +
            drawing + "'), 4326), '" +
            enteredDescription + "', '" +
            enteredUsername + "')";
        console.log(sql);
        $.post({
            url: "https://michaeldorman.carto.com/api/v2/sql",
            data: {"q": sql},
            dataType: "json",
            success: function() {
                console.log("Data saved");
            },
            error: function() {
                console.log("Problem saving the data");
            }
        });
        var newData = layer.toGeoJSON();
        newData.properties.description = enteredDescription;
        newData.properties.name = enteredUsername;
        L.geoJSON(newData, {onEachFeature: addPopup}).addTo(cartoData);
    });

    // Clear drawn items layer
    drawnItems.closePopup();
    drawnItems.clearLayers();

}
```

We will now go over the code, step by step.

The first two expressions are exactly the same as in `example-13-04.html` (Section 13.5). Again, these two expressions are used to extract the entered text in the name and description fields, as given at the precise moment when the "submit" button was clicked. The name and description values are assigned into variables named `enteredUsername` and `enteredDescription`, respectively:

```
var enteredUsername = $("#input_name").val();
var enteredDescription = $("#input_desc").val();
```

The central code block inside the **setData** function is contained inside the **.eachLayer** iteration on **drawnItems**. As shown in **example-13-02.html** (Section 13.4.1) and **example-13-04.html** (Section 13.5), using **.eachLayer** we basically apply a function on each of the layers comprising **drawnItems**. The function has a parameter named **layer**, which is assigned with the current layer in each step of the iteration:

```
drawnItems.eachLayer(function(layer) {
    // Doing something with each drawn layer
});
```

What does the internal function in the **.eachLayer** iteration do in the present case of **example-13-05.html**? Three things:

- **Construct** the **INSERT** query for adding a new record into the **beer_sheva** table
- **Send** the query to the CARTO SQL API
- **Copy** the submitted drawing to the CARTO layer, to display it on the map

Here is the code for the first part, the SQL query construction:

```
var drawing = JSON.stringify(layer.toGeoJSON().geometry);
var sql =
    "INSERT INTO beer_sheva " +
    "(the_geom, description, name) " +
    "VALUES (ST_SetSRID(ST_GeomFromGeoJSON('" +
    drawing + "'), 4326), '" +
    enteredDescription + "', '" +
    enteredUsername + "')";
console.log(sql);
```

This code section builds the SQL query for sending the currently iterated drawn shape to the database. Basically, instead of a fixed **INSERT** query, such as the one shown above (**'Point 1'** created by **'Michael'**), we are constructing the query *dynamically*, using the three variables:

- **drawing**—The GeoJSON string for the current layer, goes into the **the_geom** column
- **enteredDescription**—The description entered into the popup, goes into the **description** column
- **enteredUsername**—The name entered into the popup, goes into the **name** column

The complete query is assigned into a variable named **sql**. Using **console.log**, the value of **sql** is then printed into the console, which is helpful when inspecting our web map for potential problems.

The second part of the code takes care of sending the SQL query contained in **sql** to CARTO:

```
$.post({
    url: "https://michaeldorman.carto.com/api/v2/sql",
    data: {"q": sql},
    dataType: "json",
    success: function() {
```

```
        console.log("Data saved");
    },
    error: function() {
        console.log("Problem saving the data");
    }
});
```

The query is sent as part of an Ajax `POST` request, which is something we haven't used yet. `POST` requests are more rarely used than `GET` and a little less convenient to work with (Section 5.3.2.3). However, `POST` requests are more appropriate when *sending* data to be processed on the server, as opposed to `GET`, which is mostly used to *get* data from the server. It is important to note that the CARTO SQL API can accept both `GET` and `POST` requests, so the same request can be achieved in both ways. In this case, however, making a `POST` request is safer because the URL in `GET` requests is limited in character length. The exact lower limit depends on the browser, but can be as low as 2048 characters[17]. So, if the user has drawn a very complex geometry which results in a very long GeoJSON string, the resulting `GET` request may be rejected. In a `POST` request, the parameters are sent as part of associated *data*, rather than being part of the URL, which resolves the limitation.

To make a `POST` request we are using the `$.post` function. As discussed in Section 7.7.3, `$.post` belongs to the set of jQuery helper functions for making Ajax requests (Table 7.7). This set of functions also includes the `$.getJSON` function that we have been extensively using in Chapters 7–12. The `$.post` function, specifically, is used for making requests of type `POST`. In our case, we need to make a `POST` request to the URL of the CARTO SQL API (`"https://michaeldorman.carto.com/api/v2/sql"`), with the sent data being the `sql` string and the data we expect back from the server being `"json"`. Finally, we specify what to do when the request is successful (`success`) or when it fails (`error`). In this case, we choose to simply print either the `"Data saved"` or the `"Problem saving the data"` string in the console.

The third part of the `eachLayer` iteration, inside our `setData` function code body, transfers the drawn data to the `cartoData` layer to display it on the map without reloading the map. Basically, the drawn `layer` is translated to GeoJSON, combined with the `description` and `name` properties, then added on the map with `L.geoJSON`. Without this part, our drawing would only be sent to the database without being shown on the map, unless we reload the web page:

```
var newData = layer.toGeoJSON();
newData.properties.description = enteredDescription;
newData.properties.name = enteredUsername;
L.geoJSON(newData, {onEachFeature: addPopup}).addTo(cartoData);
```

Finally, outside of the `.eachLayer` iteration, we close the editable popup and clear the `drawnItems` feature group. The map is now ready for making a new drawing and sending it to the database.

```
drawnItems.closePopup();
drawnItems.clearLayers();
```

---

[17]https://stackoverflow.com/questions/417142/what-is-the-maximum-length-of-a-url-in-different-browsers

The complete crowdsourcing app (`example-13-05.html`) is shown in Figure 13.10.

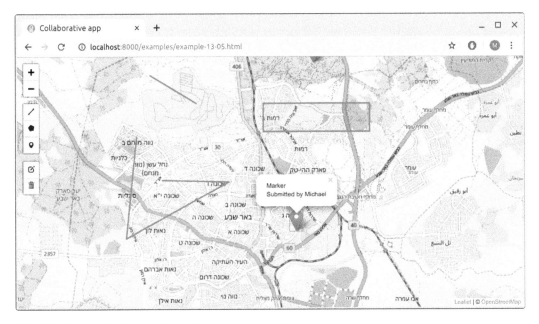

FIGURE 13.10: Screenshot of `example-13-05.html`

13.7 Exercise

- Use the previous guidelines and `example-13-05.html` to create your own crowdsourcing app:
  - Create a new table for the responses on CARTO.
  - Run the appropriate SQL query for granting `INSERT` privileges on your table (Section 13.6.1).
  - Modify the code in `example-13-05.html` (Section 13.6.2) to include your CARTO username and table name.

Part V

Appendices

A

Associated files

The following tree specifies the structure of the folder that contains the book examples, solutions, and associated files needed to run the examples, such as local copies of the required JavaScript libraries. Appendices B and C give the complete lists of the example and solution files, respectively. The entire folder is available as a ZIP file in the online version of the book (Section 0.7).

```
|-- css
    |-- images
        |-- layers-2x.png
        |-- layers.png
        |-- marker-icon-2x.png
        |-- marker-icon.png
        |-- marker-shadow.png
        |-- spritesheet-2x.png
        |-- spritesheet.png
        |-- spritesheet.svg
    |-- leaflet.css
    |-- leaflet.draw.css
|-- data
    |-- 4.5_week.geojson
    |-- county2.geojson
    |-- events.json
    |-- plants.geojson
    |-- towns.geojson
    |-- towns_pnt.geojson
|-- images
    |-- leaflet.png
    |-- marker-shadow.png
    |-- north.svg
    |-- redIcon.png
    |-- south.svg
|-- js
    |-- jquery.js
    |-- leaflet.draw.js
    |-- leaflet.heat.js
    |-- leaflet.js
    |-- turf.js
|-- example-01-01.html
|-- example-01-02.html
|-- .................(more examples)
|-- example-13-05.html
|-- example-13-05-s.html
```

```
|-- map.js
|-- solution-03.html
|-- solution-04.html
|-- ...................(more solutions)
|-- solution-11-s.html
|-- solution-12.html
```

B

List of examples

Table B.1 lists all of the book's code examples, in order of their appearance. Note that alternative versions of examples that use a custom SQL API server, instead of CARTO, have the `-s` suffix in their name (e.g., `example-09-01-s.html`) and are marked with (**S**) (see Section 9.2.2).

TABLE B.1: List of examples

Chapter	Example	Description
Chapter 1	example-01-01.html	A minimal web page
	example-01-02.html	White space collapsing
	example-01-03.html	Lists
	example-01-04.html	Images
	example-01-05.html	Tables
	example-01-06.html	Input elements
Chapter 2	example-02-01.html	CSS conflicts
	example-02-02.html	CSS colors
	example-02-03.html	CSS fonts
	example-02-04.html	CSS box size
	example-02-05.html	CSS box position
	example-02-06.html	Hurricane scale
	example-02-07.html	Hurricane scale with CSS
	example-02-08.html	Map description text
	example-02-09.html	Map description positioned
	example-02-10.html	Map description customized
	example-02-11.html	Map description on web map
Chapter 4	example-04-01.html	Hello JavaScript
	example-04-02.html	Earth poles
	example-04-03.html	jQuery operating on selection
	example-04-04.html	jQuery event listener
	example-04-05.html	Hello JavaScript (jQuery)
	example-04-06.html	Earth poles (jQuery)
	example-04-07.html	Event object
	example-04-08.html	Populating list
	example-04-09.html	jQuery content from data
Chapter 6	example-06-01.html	Vector tiles (Mapbox GL JS)
	example-06-02.html	Basic map
	example-06-03.html	Adding marker
	example-06-04.html	Adding line
	example-06-05.html	Adding polygon
	example-06-06.html	Adding popup
	example-06-07.html	Adding map description

C

List of exercise solutions

Table C.1 lists solutions to the exercises that appear in the end of each Chapter. Note that alternative versions of the solutions that use a custom SQL API server, instead of CARTO, have the `-s` suffix in their name (e.g., `solution-09-s.html`) and are marked with (**S**) (see Section 9.2.2).

TABLE C.1: List of exercise solutions

Chapter	Solution	Description
Chapter 3	solution-03.html	Segment coordinates
Chapter 4	solution-04.html	Expanded calculator
Chapter 6	solution-06.html	Travel plan map
Chapter 7	solution-07.html	EONET real-time events
Chapter 8	solution-08.html	Towns circle markers
Chapter 9	solution-09.html	Species list
	solution-09-s.html	Species list (**S**)
Chapter 10	solution-10.html	Genus/species dropdown menu
	solution-10-s.html	Genus/species dropdown menu (**S**)
Chapter 11	solution-11.html	Distance to nearest plants
	solution-11-s.html	Distance to nearest plants (**S**)
Chapter 12	solution-12.html	Draggable Great Circle

Bibliography

Bassett, L. (2015). *Introduction to JavaScript Object Notation: a to-the-point guide to JSON.* O'Reilly Media, Inc, Sebastopol, CA, USA.

Crickard III, P. (2014). *Leaflet. js Essentials.* Packt Publishing Ltd, Birmingham, UK.

Crockford, D. (2008). *JavaScript: The Good Parts: The Good Parts.* O'Reilly Media, Inc.

DeBarros, A. (2018). *Practical SQL: A Beginner's Guide to Storytelling with Data.* No Starch Press, San Francisco, CA, USA.

Dent, B. D., Torguson, J. S., and Hodler, T. W. (2008). *Cartography: Thematic map design.* WCB/McGraw-Hill, Boston, 6th edition, New York, NY, USA.

Dincer, A. and Uraz, B. (2013). *Google Maps JavaScript API Cookbook.* Packt Publishing Ltd, Birmingham, UK.

Farkas, G. (2016). *Mastering OpenLayers 3.* Packt Publishing Ltd, Birmingham, UK.

Gratier, T., Spencer, P., and Hazzard, E. (2015). *OpenLayers 3: Beginner's Guide.* Packt Publishing Ltd, Birmingham, UK.

Langley, P. J. and Perez, A. S. (2016). *OpenLayers 3. x Cookbook.* Packt Publishing Ltd, 2nd edition, Birmingham, UK.

Murray, S. (2017). *Interactive Data Visualization for the Web: An Introduction to Designing with D3.* O'Reilly Media, Inc, 2nd edition, Sebastopol, CA, USA.

Murrell, P. (2009). *Introduction to Data Technologies.* Chapman and Hall/CRC, Boca Raton, FL, USA.

Neuwirth, E. (2014). *RColorBrewer: ColorBrewer Palettes.* R package version 1.1-2.

Newton, T. and Villarreal, O. (2014). *Learning D3.js Mapping.* Packt Publishing Ltd, Birmingham, UK.

Nield, T. (2016). *Getting Started with SQL: A Hands-On Approach for Beginners.* O'Reilly Media, Inc., Sebastopol, CA, USA.

Obe, R. and Hsu, L. (2015). *PostGIS in Action.* Manning Publications Co., 2nd edition, Shelter Island, NY, USA.

Pebesma, E. (2018). Simple Features for R: Standardized Support for Spatial Vector Data. *The R Journal*, 10(1):439-446.

QGIS Development Team (2018). *QGIS Geographic Information System.* Open Source Geospatial Foundation.

R Core Team (2018). *R: A Language and Environment for Statistical Computing.* R Foundation for Statistical Computing, Vienna, Austria.

Rubalcava, R. (2015). *ArcGIS Web Development.* Manning Publications, Co., Shelter Island, NY, USA.

South, A. (2011). rworldmap: A New R package for Mapping Global Data. *The R Journal,* 3(1):35–43.

Tufte, E. (2001). *The quantitative display of information.* Graphics Press, 2nd edition. Cheshire, CT, USA.

Wickham, H. (2018). *nycflights13: Flights that Departed NYC in 2013.* R package version 1.0.0.

Wilke, C. O. (2019). *Fundamentals of Data Visualization: A Primer on Making Informative and Compelling Figures.* O'Reilly Media, Inc., Sebastopol, CA, USA.

Xie, Y. (2016). *bookdown: Authoring Books and Technical Documents with R Markdown.* Chapman and Hall/CRC, Boca Raton, FL, USA.

Xie, Y. (2018). *bookdown: Authoring Books and Technical Documents with R Markdown.* R package version 0.9.

Index

Milton Keynes UK
Ingram Content Group UK Ltd.
UKHW050309111024
449327UK00049B/366